INSIDE THE WELL

INSIDE THE WELL

The Midland, Texas Rescue
of Baby Jessica

LANCE LUNSFORD

TEXAS TECH UNIVERSITY PRESS

Copyright © 2024 by Lance Lunsford

All rights reserved. No portion of this book may be reproduced in any form or by any means, including electronic storage and retrieval systems, except by explicit prior written permission of the publisher. Brief passages excerpted for review and critical purposes are excepted.

This book is typeset in EB Garamond. The paper used in this book meets the minimum requirements of ANSI/NISO Z39.48-1992 (R1997). ♾

Designed by Hannah Gaskamp
Cover design by Hannah Gaskamp
Cover photo by Scott Shaw, used with permission from the *Odessa American*

Library of Congress Cataloging-in-Publication Data

Names: Lunsford, Lance, 1977– author. Title: Inside the Well: The Midland, Texas Rescue of Baby Jessica / Lance Lunsford. Description: Lubbock: Texas Tech University Press, 2024. | Includes index. | Summary: "A firsthand account of the 1987 media-charged rescue of Baby Jessica McClure, with a chronicle of the community aftermath once the news cameras left"—Provided by publisher.
Identifiers: LCCN 2023058473 (print) | LCCN 2023058474 (ebook) | ISBN 978-1-68283-218-9 (paperback) | ISBN 978-1-68283-219-6 (ebook)
Subjects: LCSH: McClure, Jessica, 1986– | Children's accidents—Press coverage. | Search and rescue operations—Press coverage. | Midland (Tex.)—Press coverage.
Classification: LCC HV670.U52 L66 2024 (print) | LCC HV670.U52 (ebook) | DDC 363.13/81—dc23/eng/20240311
LC record available at https://lccn.loc.gov/2023058473
LC ebook record available at https://lccn.loc.gov/2023058474

24 25 26 27 28 29 30 31 32 / 9 8 7 6 5 4 3 2 1

Texas Tech University Press
Box 41037
Lubbock, Texas 79409-1037 USA
800.832.4042
ttup@ttu.edu
www.ttupress.org

To my cherished friends and family, for the unwavering support and willingness to hear out every newly discovered detail in the stories that went into this book. Their attention and feedback were instrumental in stitching together an intricate narrative from what would have otherwise remained fragmented and disparate elements. They helped form a cohesive tale of West Texas that will certainly once again resonate with a global audience.

Especially to my four girls, Michelle, Jovi, Lucy, and Daphne.

FAME is a food that dead men eat,
I have no stomach for such meat.
In little light and narrow room,
They eat it in the silent tomb,
With no kind voice of comrade near
To bid the banquet be of cheer.

But Friendship is a nobler thing,
Of Friendship it is good to sing.
For truly, when a man shall end,
He lives in memory of his friend,
Who doth his better part recall,
And of his faults make funeral.

—HENRY AUSTIN DOBSON

CONTENTS

	PREFACE	XI
	Introduction	3
CHAPTER 1:	"Those People Are Going to Need Some Help for a Long Time" ROBERT O'DONNELL, FORMER FIREFIGHTER	7
CHAPTER 2:	Officers Descend on Tanner Drive	11
CHAPTER 3:	The Arrival	31
CHAPTER 4:	Life Unraveled MOMENTS AFTER THE OKLAHOMA CITY BOMBING	41
CHAPTER 5:	Doing It Live in the Permian Basin	51
CHAPTER 6:	Too Far Gone	75
CHAPTER 7:	Oil Town Good ol' Boys Come to the Rescue PROSPECTS FOR QUICK RESCUE SHINE BRIGHT	81
CHAPTER 8:	Fifty-Eight Hours of Daylight NIGHT AND DAY COLLIDE AT THE RESCUE SITE	89
CHAPTER 9:	Paula, Get Your Gun FIFTEEN YEARS AFTER THE RESCUE	103

CONTENTS

CHAPTER 10: Good Gawd!   131
OIL COUNTRY, SPIRITUALITY, AND A FAITH IN SURVIVAL

CHAPTER 11: The Last Hours of Robert O'Donnell   139

CHAPTER 12: Day Two Nears End with Hope for a Breakthrough   143

CHAPTER 13: Jessica McClure Proves Out CNN's Bet on Cable Model   151
THE NEWS BUSINESS GOES LIVE

CHAPTER 14: Cops Knock on the Cop's Door   167

CHAPTER 15: The Walls Come Tumbling Down   173

CHAPTER 16: What Happens Next?   189
THE AFTERMATH

CHAPTER 17: Oprah Comes to Town   197

CHAPTER 18: A Hospital under Siege   211

CHAPTER 19: "I Know How I'm Gonna Do It"   221

CHAPTER 20: Boys Will Be Boys   225
COLLEAGUES TURN ON O'DONNELL AMID THE SPOTLIGHT

CHAPTER 21: As Fast as He Rose, He Fell   231

CHAPTER 22: "I Don't Know Why I Did What I Did"   237

Epilogue   241

ACKNOWLEDGMENTS   253
REFERENCES   257
INDEX   259

PREFACE

We Had Big Plans on October 14, 1987

AT THE END OF OUR STREET, THE 4700 BLOCK OF Preston Drive on the western edge of Midland, a large empty field sat at the precipice of what early West Texas explorers called "The Big Empty." To me, nine years old at the time, it was the edge of the Earth—pure, endless possibility. My friends and I would spend hours exploring the area over rutted dirt roads and winding paths that were likely old cattle trails cutting through shoulder-high thickets of wiry mesquite. We would sometimes venture to the field's far edges on dirt bikes with a feeling of adventure like the first Apollo astronauts must have felt preparing to reach beyond the known boundaries of humankind.

There were few rules out there. We took advantage of the freedom. Oncoming cars could be seen with enough time to prevent us from getting caught red-handed with some of those innocent preteen explorations like busting empty glass beer bottles on an abandoned section of concrete culvert or experimenting with a friend's BB gun on a series of targets fashioned from whatever random objects might have been lying around. Occasionally, the freedom and adventure came with a grim dose of reality, like on one cold Saturday morning when we came upon the remains of a rotting coyote corpse. We puzzled over its snarled teeth framed by thin curled lips and snout—its face in paralyzed expression of offense from its final moments of fight. Its paws were inexplicably bound tightly with thin wire.

Our plans for that Wednesday afternoon in mid-October included the initial preparations for the creation of an underground fort out in the field. We had chosen a spot set back in the brush that was tall enough to hide our whereabouts. It would eventually be overtaken by varmints and by what we suspected, but wisely never verified, was a family of rattlesnakes. That was still months off, however. That Wednesday was about possibility and preparation, and our little crew of kids from Preston Drive spirited what tools we could from our dads' garages and sheds to begin construction.

We rocketed off our school bus ready to make the one-mile trek by dirt bike home from the drop-off at Rusk Elementary School. Instead of making a move toward the school's bike rack, however, I was met unexpectedly by my younger brother. He had been given instructions that would disrupt our crew's activities: we were to meet our mom in the school pickup line. He said we would not be heading home but instead to our grandparents' house, which would certainly mean an end to the afternoon's secret construction plans.

I was incensed—my scheduled construction co-opted with no possibility for a counterargument.

"I'm not sure why," my brother, Donnie, asserted. "A baby fell in a well by Grandma and Grandaddy's house."

That is a lot for a fourth grader to take in.

While wondering what exactly this baby had to do with me and how my family's presence might somehow be required, I also pictured what every other human being on the planet pictured when they heard the news for the first time about "Baby Jessica" McClure and an abandoned water well in a backyard in Midland, Texas: the colorful fairy-tale portrait of a wishing well with a makeshift mechanical winch to drop and retrieve a tin pale.

As we arrived at Tanner Drive in my family's white Ford Aerostar minivan, my mom, Ann DeLong Lunsford, tried to turn left onto the street but was rebuffed by a Midland police officer directing traffic away from Tanner. "We're trying to get to my in-laws' house right there," mom politely protested, pointing to a house just three houses away on the western side of the street.

The officer soon learned what I already knew about dealings with Ann Lunsford. She was not to be trifled with. She had been well schooled

PREFACE

in the dark arts of West Texas haggling. Having grown up along a low-income stretch of industrial highway in a house her dad built himself before she was born, her skills came from years of experience negotiating over things as small as the fat-to-protein ratio of a selection of bacon with the butcher (they knew each other by name) or the ripeness of the latest batch of heirloom tomatoes with the store manager and as big as the interest rate on her home mortgage with her bank's loan manager. Big or small, she took them all on with gusto.

She petitioned the officer politely once more, and he caved—as I knew he would—waving us through the entrance to Tanner now clogged with police and emergency response vehicles.

We wasted no time in making our way through my grandparents' house to their alley, where we easily navigated through the fence-lined thoroughfare to a crowd of emergency workers and a giant hulking green rig with moving chain gears. Along a fence separating us from where the rescue was unfolding, I recognized a reporter from one of the local television news affiliates. He was talking to himself, a reporter's notebook in his hands. Despite the chaos of the scene, I zeroed in on what might look like a randomly pacing mad person incoherently talking to himself, only to realize he was memorizing his forthcoming report from the scene to be broadcast in real time over the air. The magnitude of that moment hit me. And stuck. Knowing that what was happening on this small patch of ground in a low-income neighborhood in a space that I shared with police, firefighters, EMTs (emergency medical technicians), and journalists would be broadcast across the city and beyond was incredible. I was fascinated to be standing at the center of a major event, witnessing something from which the images would be split into tiny electrons, sent over the airwaves, and put back together again, line by line, into television screens far away. Later I would learn that these emotions were not unique. Nor were they immature. Biographies from journalism greats as different as Walter Cronkite and Hunter S. Thompson have chronicled similar brushes with early impactful events and feelings.

My brother and I made our way to the six-foot fence blocking our view of the rescue. I found a toehold in the wood fence frame and hoisted myself high enough to clear the top and establish a line of sight directly to the wellhead, where men were lying on the ground and crouched over a small rusty pipe. Clearly not a colorful wishing well.

From those initial hours immediately following Jessica's fall into the well, we went on about our business with school and work as the rescue efforts continued. After school the next day, on Thursday, we returned to Tanner Drive and the rescue site as more media poured into town. Interviews with the *Dallas Morning News* took place in my grandparents' living room. A reporter from a Hawaiian newspaper asked about a recent interaction with neighbors in the area. Emergency workers and other reporters used their bathroom. The eyes of the world were on our little town, and we were in the thick of it.

Three days later, we watched from my grandparents' living room as the live television broadcast continued from three houses down, and dusk turned to night. As the rescue's final moments unfolded, we sprinted out the back door as the ambulance carrying "Baby Jessica" McClure made its way from what had been the center of the universe for three days, in a first-of-its-kind moment in a world that did not yet know anything about going viral or what a social media influencer would be.

It would be fifteen years later that I would make the questionable career decision after graduating from Texas A&M with a degree in journalism to start work as a police and courts beat reporter for the *Midland Reporter-Telegram*. Few, if any, recent college graduates were coming to Midland in 2001 to start their careers—not in the way the oil patch would eventually attract new graduates with valuable offers from the world's largest oil firms or the industry's most innovative and adventurous investors. Midland was still reeling from the oil bust, and many around town no longer considered it a boom town. It had not yet experienced the return of the $100 per barrel oil that would eventually incentivize the deployment of new technologies that could make profitable the retrieval of oil once considered unreachable—or not yet worth the struggle to break it free from composite oil shale.

Somewhat begrudgingly, at first, I took my editor's assignment to write up the fifteen-year anniversary story of the Baby Jessica rescue.

PREFACE

The previous sixteen months had been difficult. I had taken on reporting tasks that challenged city administrations, had talked with parents of recently deceased children who perished in drownings and car accidents, and had listened to court proceedings for violent murders and rapes. But there were upsides to my decision to start my career in my hometown. I spent a lot of time with firefighters, doing overnight ride-alongs with their crews and showing them I was not afraid to get my reporter's hands dirty on long shifts, getting close to fiery car wrecks and fires. After 9/11, some had begun talking about what was then a little-known issue affecting firefighters—post-traumatic stress disorder (PTSD).

I was introduced to Vaughn Donaldson, a Midland Fire Department (MFD) district director, who, following his own concerns in the aftermath of the rescue of Jessica McClure, earned a bachelor's degree in psychology at the University of Texas–Permian Basin. We spent hours talking about his research and his experiences with fellow firefighters.

As I made my way back to Tanner Drive, where I had spent hundreds of days of my youth at my grandparents' house, I knocked on the door of Maxine Sprague, who immediately recognized my name and invited me in to visit. Maxine and her husband lived next door to the home where Jessica had fallen into the well. As we chatted, Maxine shared her vivid recollections of the event, and I felt a nostalgia and familiarity as she called back to her husband telling him that one of Dutch and Margie Lunsford's grandsons was there to chat and was now working for the *Reporter-Telegram*.

When I left the Spragues' home, I began to see the story of Baby Jessica not just as the rescue of an eighteen-month-old infant. It was not about what she might remember as a teen. It was not about her mom and dad. And it was not about my experience of it as a rambunctious nine-year-old. Instead, it was a unique story about West Texas and a community's shared determination in the face of a challenge and amid economic failure and dashed hopes and dreams.

I quickly reconnected with Donaldson, who helped shed light on the darkness that followed Midland firefighter Robert O'Donnell in the years after the rescue. Those conversations helped round out a separate sidebar story to the series that would narrow in on O'Donnell's experiences as a microcosm for the much more universal experience of missing the heat of the limelight when it fades and the lengths to which some will go to keep it. I called on those memories from that first day of the rescue I witnessed from my grandparents' backyard, seeing the media arrive on the scene. I reached out to the television affiliates competing with my own newspaper and interviewed their news directors and managers to fold their perspectives into the story and begin to define the phenomena that created the first major twenty-four-hour news cycle event of our time. Without hesitation, Rick Wood and Dave Foster at the NBC affiliate agreed to my in-depth interviews, providing personal recollections and invaluable insight.

After turning in the two-day series of stories along with the O'Donnell sidebar to my editor, I received rare kudos for the angle I took on O'Donnell from *Reporter-Telegram* city editor Gary Ott. "That was a good story. I hadn't heard that about him [O'Donnell] before. . . . I just figured he had other issues and problems related to drugs and alcohol or something."

From there, this book was born. Even Midland and those closest to the story of Jessica's rescue did not have a full accounting of what Robert O'Donnell experienced—his immediate exposure to the limelight, his dealings with jealous coworkers, his brushes with celebrity and celebrities, his likeness warped in Hollywood's version of his real identity.

Although a significant profile of O'Donnell's story by Lisa Belkin appeared in the *New York Times Magazine* in July 1995, I questioned whether there was more to the story and whether there might be nuances in a deeper telling that would resonate with a wider audience who would see the larger themes of PTSD, the toll of dealing with limelight from new media and technology, and audience behavior.

Just as important is understanding how CNN strategically disrupted the business of reporting the news; that understanding is critical, especially today. Enabled by technology and the government's investment in satellite communications, the cable business birthed a subscription-based funding model that depended less on advertising inventory than did network television. Strangled by time allocated to advertising, the three powerful network broadcast corporations did not have the leeway that CNN had to fulfill audiences' desires to focus, transfixed and without advertising breaks, on sensational live news events. The popular view is that CNN's innovation was in the delivery of live news with on-scene coverage and the excitement that comes with it. In fact, CNN's innovation was their use of cable subscriptions, which untethered their live TV news from the financial constraints that bound competing over-the-air networks to advertisers. CNN was free to stay live on a story without ever withdrawing a viewer's attention from the scene of a breaking news event. Traditional networks, by contrast, were always under pressure to interrupt live broadcasting with the sponsored commercial breaks of their paying customers. Without CNN's innovative finance model, Baby Jessica's rescue might have been yet another major news event in history and not nearly the worldwide phenomenon it turned into.

Today, approaching forty years on, with a news cycle more frenetic than the one CNN pioneered and with a nearly insatiable appetite for all things celebrity, O'Donnell's and Midland's story is even more relevant, offering a timeless lesson in human ambition and frailty.

INSIDE THE WELL

INTRODUCTION

Seven Years after the Rescue

IT FELL TO DR. JERRY D. SPENCER TO OFFICIALLY declare the cause of Robert O'Donnell's death on April 24, 1995. Training in medicine was hardly necessary. There were plenty of clues, including a combination of prescribed medications filtered from O'Donnell's blood. Alcohol was not a factor. Not a drop was found among the antidepressants, sleeping aids, and painkillers.

> Amitriptyline, a prescribed antidepressant.
> Butalbital, a barbiturate prescribed for ailments like back pain.
> Valium or perhaps Flurazepam, a sleeping pill.
> No trace of alcohol.

That booze played no role came as a surprise to those who heard about O'Donnell's demise. Most of those people had done little to pry into his story, making assumptions along the way that included speculation of frenzied alcoholism following his dance with fame. It was an easy way, perhaps, to explain away the years of tumult that resulted in O'Donnell finding no other resort. The stockpile of rumor and innuendo around O'Donnell's alcohol consumption caused enough friction in his life to fuel a mile-wide West Texas grassfire, the kind that regularly blew white smoke into the clear blue sky for hundreds of miles. It made sense only to one of his fellow firefighters, Vaughn Donaldson, who had delved deep into understanding the details of what would eventually become known as PTSD, post-traumatic stress disorder; but it was at the time a little-understood condition and rarely, if ever, taken seriously.

Donaldson, an administrator for the MFD, had devoted himself to learning more about the subject after seeing how O'Donnell responded to the aftermath of the Jessica McClure rescue and its residual media limelight.

Still, it was not the chemical cocktail running through his bloodstream that brought O'Donnell's body to lie, cold and blue, on a metal table for Dr. Spencer to examine. Those who knew O'Donnell best in this small Texas city knew what had led to O'Donnell's demise, including O'Donnell's calf-roping, hard-living brother, Ricky. Each person who knew O'Donnell had a painful take on the troubled path leading to where Ricky would find his brother's body slumped over the steering wheel of his pickup truck on the family's ranch.

"Ever since that Jessica deal, his life fell apart," Ricky told a reporter, whose story ended up in print half a world away in Buffalo, New York.

Ricky had been the first to find Robert after combing the desert ranch in the dead of night. The family and others had searched for hours with no luck.

Robert O'Donnell was not supposed to be lying here in the morgue, his body exposed, on full display to reveal whatever secrets it had been hiding. It was up to the coroner to figure out how someone who had once been beloved and heralded had ended his own life. Dr. Spencer hovered over the man who was once called on from all over the world to speak as an authority on teamwork, commitment, and doing what it takes when duty calls. O'Donnell had been known for his lean, muscular build. Previously, O'Donnell was seen as a fit athletic figure. He had once had a somewhat authoritative posture, like a police officer, with a shy but confident poise; now O'Donnell's body lay stiff, its fluids drained in preparation for interment following Dr. Spencer's examination.

How had a life with such promise, filled with accolades for an achievement known worldwide, come to this? Even after a full accounting of his death, a complete examination seems impossible, and to those who continue to ponder it, it remains a mystery.

The cause of death was quite clear, if not blatantly apparent.

The roof of O'Donnell's mouth had been torn to shreds, a black hole left by tiny pellets, scattered from the exploding barrel of a single-shot

.410 bore shotgun. Although such a blast could easily explain what Dr. Spencer was left to examine, the true cause of death of this husband, father, brother, son, and hero did not need a doctor. Everyone knew, but no fingers were pointed. For those aware of the details, it was simple. Dr. Spencer would not be needed. His toxicology, useless. Spencer's experience in medicine could explain it no more clearly than any old buddy who knew O'Donnell. It was clear. It was sad, and it was damned hard to admit it or say it out loud.

One writer-turned-editor termed it as a death by toddler—a little girl who, indirectly and with no fault of her own, had the misfortune of falling into an abandoned water well on October 14, 1987. For three fall days, the full weight of the world's attention bore down on a nine hundred square-foot section of Midland, Texas, a boom-and-bust oil town. O'Donnell was just a young dad, a husband and firefighter who seemed to have found himself a place in the world as a paramedic for the MFD.

For Robert O'Donnell, the day of that little girl's rescue from the bottom of a dark well shaft was the beginning of the end. For the dozens of others involved in the rescue that created a worldwide media frenzy, it was a life-changing experience. In a series of events that followed, O'Donnell came to find himself on a collision course with his own purpose and existence.

CHAPTER 1

"THOSE PEOPLE ARE GOING TO NEED SOME HELP FOR A LONG TIME"

Robert O'Donnell, Former Firefighter

IT WAS SUNDAY NIGHT, AND ROBERT O'DONNELL WAS IN the deepest string of deep funks. The firefighter-paramedic turned asbestos-removal technician had failed at everything he had done of late. Everything, that is, but being a hero. From that lofty perch there was no place to go. No place but down. And, he had done that too.

On this night, he was slumped low in a chair at the home of David Poe, his stepfather, who twenty years earlier had married his mother in a gaudy Las Vegas chapel. At thirty-seven, O'Donnell was now on his third career. His second—as a paramedic and a fireman—had brought him the greatest rewards. It was the one he was best at, but he had come to the conclusion that he could no longer make things work in the profession of a firefighter-paramedic. His employer agreed.

Hoping his life had finally turned around, O'Donnell was three days into a new career. He had taken on a new role with Lubbock's King Consultants. King was a low-profile company specializing in the removal of asbestos from 1940s- and 1950s-era buildings, those constructed during the days when the fireproof insulation was touted for its safety and durability and before its cancerous health effects were fully known.

His fellow employees at King hardly knew him. He was just another face—a new hire learning the ropes. Few, even among his supervisors, knew of his previous heroics. Uncomfortably recalling O'Donnell, one King administrator later admitted to not knowing anything about O'Donnell and his experience as the rescuer of Jessica McClure.

"We didn't know anything. Anything about him. He had hardly worked here," the administrator noted dismissively. He said the words as someone might who was trying to rid his company of a quick black footnote. By the time O'Donnell had come to work as an asbestos-removal technician, he was almost eight years removed from that balmy night in Midland when he had descended a rocky hole to save a little girl's life, becoming for a moment a national superstar.

On this night of April 23, 1995, O'Donnell could see the end of the line.

Television was not helping his mood much either. From the set at his mother's home, horrific images spilled into the living room. Three days earlier, a military veteran and malcontent had declared war on his own country, detonating seven thousand pounds of ammonium nitrate fertilizer in a rented Ryder truck in front of the Alfred P. Murrah Federal Building in Oklahoma City—taking 168 lives in a flash.

The explosion had ripped through seven floors of reinforced concrete, cutting a massive hole into the side of the building housing federal offices for 550 employees who worked for agencies like the Social Security Administration, the US Department of Housing and Urban Development, and the US Secret Service. On the first floor, a daycare operated for employees in the building. The death and carnage spewing from the television in front of O'Donnell, sitting in his mother's living room, only sent his mood to darker recesses.

O'Donnell was in search of some rest. He had been between jobs for too long before he got the asbestos abatement role in Lubbock. He was years removed and ostracized from a career he loved and worked so hard for. Estranged from his wife and children, with only occasional visits, his distance from a reality he had once loved was overwhelming. The world he knew had been rushing away from him for a while. His possessions, his stability, the life he had built had all been sucked away in a vacuum.

"THOSE PEOPLE ARE GOING TO NEED SOME HELP FOR A LONG TIME"

What once had burned bright, his heroism, had faded to embers. As the television's noise hummed with news of the rescue of those who might still be trapped in the rubble of the Murrah Building, news crews descended on the ruins as firefighters, EMS (emergency medical services) personnel, and police pulled bodies from the wreckage. Pointing to the collection of images flickering from the television set, O'Donnell's mom, Yvonne, later recalled Robert's words: "Those people are going to need some help for a long time."

O'Donnell had been speaking not of the rescued—but of the rescuers. As he sat and watched news coverage, smoke still billowing, Oklahoma City police sergeant Terrance Yeakey emerged from among others and stepped to the forefront of the rescue site. Pulling four people from the wreckage, he soon after fell through two floors of the building and injured his back. Firefighter Chris Fields also emerged carrying a young toddler who showed no signs of life; a news photographer captured the devastating moment for public consumption.

CHAPTER 2

OFFICERS DESCEND ON TANNER DRIVE

WHILE THE SOMEWHAT UNASSUMING, QUIET TOWN OF Midland knew its share of shady and even criminal activity, most of that took place quietly and tucked away, making room for its reputation as the town of the buttoned-down corporate oil executives, bankers, geologists, and engineers. The real reputation for hell-raising would be left for its less respectable neighbor: Odessa. Known more as a base for the working-class culture that fueled the Permian Basin's oil and gas industry, Odessa was about the same size as Midland but was left to the roustabouts—the oil field roughnecks and the service industry workhorses. Midland's sister city was considered a black-sheep town. Midland, by contrast, expressed itself as a solidly corporate oil town where deals were made and business strategies were devised. City folk in Midland considered themselves to be of the business and political class; it was a place where million-dollar deals were made over lunch, and legislators raised donations at the downtown Petroleum Club.

In a 1975 article in *Texas Monthly*, Texas playwright Larry L. King—drawing from colorful and legendary Texas lawyer Warren Burnett—put it this way: "A long time ago, Midland took the high road. That's why the law offices over there [in Midland] have drapes and deep rugs, and everybody talks in hushed tones. Odessa, on the other hand, has little demand for corporate lawyers. We are a bunch of sweat-hog lawyers getting it on in the courtrooms. We tend to drink in bars. Midland lawyers drink at home. Odessa, in particular, shares instincts with Fort Worth; Midland is more of Dallas, serious minded and buttoned-down."

Burnett's and King's assessments of the cities were not isolated. Citizens of the two communities viewed each other with a crooked brow transfixed in an expression like an elder's wordless scorn shaped by a lifelong distrust based in misdeeds from decades gone by. Replacing what otherwise should have been a neighborly rapport instead was a not-so-quiet simmering resentment.

Odessans saw the swagger of their Midland neighbors as a kind of preening and swanning—window dressing for a world of make-believe. Odessans would mock Midlanders, finding humor in their self-aggrandizing "upper class" aspirations despite earning only average wages. "Midland's full of eight-thousand-dollar-a-year millionaires," Odessans would quip. Midlanders, in turn, indicted Odessa as a place to raise hell, while Midland was cast as a place to raise a family.

The people in these barely separated communities could not even identify their own shared likenesses for having landed here in the middle of nowhere—and seemingly for no reason. The first settlers in the community did not plant themselves here with the cornerstone of a natural resource. In many other early Texas settlements, ranchers would stock their herds along a riverbank before setting up shop and diversifying into mercantile exchanges to barter goods for newly arriving settlers. In 1849, a little more than a decade after Texas fought for and won its independence from Mexico, Captain Randolph B. Marcy identified viable water suitable for cattle ranching amid what was deemed to be nutritious grasslands. His West Texas expedition had been funded by taxpayers as an exploration that would eventually help build a defense against the Indian presence that threatened settlers. Soon after, the expedition's path forged a plan for a railway that would support the cattle infrastructure and define this region. Settlers would thus create a ranching industry that thrived for more than a century. When explorers later surveyed the land, with their hopes set high on valuable grasses and water, they were disappointed and confused by what they found. Marcy, although known for his achievements in westward expansion and exploration, had apparently explored pooled waters that had later evaporated and were now gone.

Nevertheless, settlers came to the area with great expectations for creating a new life based on farming and ranching. When those early settlers arrived, several put down roots so firmly that even the dust storms that occasionally plagued the plains could not dissuade them from persevering. They entrenched themselves, built families, and found ways to survive the summer heat and the cold, arid winters.

The power of Midland—as the economic epicenter of the West Texas Permian Basin—grew immensely, but not in the same manner as great cities like Chicago, New York, Houston, and Dallas. The prowess of the small desert city is more about density than it is about frequency. Its wealth per capita from the 1950s to 1980s became part of the city's lore; it was, supposedly, higher than that of any other city in the world. Profiles of the oil and gas economy include repeated references to the presence, for a time, of a Rolls Royce dealership located along Highway 80 between Midland and Odessa. Little record of it exists, but it's referenced so often, it is not questioned. Writers for international newspapers, *Texas Monthly*, and National Public Radio highlight it as part of the region's character. *New Yorker* journalist Susan Orlean, in a 2000 profile of then-governor George W. Bush's hometown, mentions the dealership as shorthand measurement of the boom's good times and of concentrated wealth. Orlean writes, "People in Midland take in huge amounts of money, they lose huge amounts of money—then they move on to the next day. It's a manic depressive city, spending lavishly and then desperately suffering."

More importantly, though, this per capita wealth was hardly half a century old. The rapidity of the onset of this wealth is staggering when measured against the area's relative youth as a settled land. The historic oil barons of Texas emerged from the East Texas oil fields in the earliest part of the twentieth century. But not until the 1930s–1960s did Midland's oil wildcatters start to emerge. Oil plays were found in West Texas in a geographic anomaly that had trapped ancient sea life in a basin when the oceans eventually receded. This area is known as the Permian Basin.

The resulting crude would certainly become part of the Texas oil narrative, fueling its worldwide influence and, by chance, becoming a crucible of power and prestige on the world stage. Oil would soon transform the entire state more rapidly—and more permanently—than any other industry had impacted any other region in the country.

The oil production levels from the mid-1900s were a ghostly memory in October 1987. The full brunt of an oil crash had come down hard on Midland, and the rest of the Permian Basin felt more than a pinch. Complete and total economic collapse had set in on the area. Texas as a state, too, felt the hurt. In 1986, the *Houston Chronicle* profiled Alan Hutchinson, the former president of a Houston-based oil exploration firm. A photo of forty-six-year-old Hutchinson prominently features him waiting tables at an upscale restaurant called Mimi's. Like him, once towering figures in the oil field littered the landscape of Texas. Some tried to hang on in West Texas in the hope of another boom. They went about their lives, raising their families and attempting to maintain their normal everyday patterns. Looking for employment. Going to work in other careers. Taking their kids to school. As one of them put it three decades later, "I was determined to get back what I once built up and worked hard for, but I wasn't going to sit around and worry about it. I knew I could get it back."

For many of them, getting back on track would mean conforming to the tightly configured template that followed World War II's end. Families soldiered on, enrolling in Boy Scout troops and signing up for the YMCA soccer league. They looked for ways to expand their horizons and prepare college investments and retirement funds.

Wednesday, October 14, was one such normal day in the fall of 1987, when students in Miss Hobbs's fourth grade homeroom at Lamar Elementary School wrestled their way through another math assignment—a worksheet of long division problems left blank and waiting for answers. On this side of Midland's south-central core, a police siren or two during the day would bring little or no alarm.

Students at Lamar Elementary were a mix of children from across Midland. Bused from every corner of the city, some students were part of the upper middle class. Some were part of the lowest end of the socio-economic spectrum. Hispanics mingled with white students, as did Black students. They ate together at lunch tables and did little to reflect the harmful prejudices that had segregated their community's school a little more than a generation before. It had taken a court order and a complex school busing scheme to get this kind of integrated community. Wedged in a tidy corner of Kessler Avenue, a quick eastward turn from the major thoroughfare of Midkiff Drive, Lamar Elementary sits in a quiet neighborhood of slowly aging homes.

When a series of sirens on the morning of October 14, 1987, whirred their way to life and kept coming, growing in strength and volume, several of the students finally looked up from their work. Puzzled by what could have necessitated this kind of sustained alarm, they sat for a moment, adrift in the noise.

Less than a mile away, the curtain on a mother's nightmare had parted. In a backyard where she babysat for an unlicensed home daycare, Reba McClure, known as "Cissy," was trying to figure out how the tiny voice of her toddler daughter could be coming from the dark, black emptiness of the eight-inch-diameter well casing.

The well casing, which from the surface looked like nothing more than a brown rust-colored pipe, sticking several inches out of the ground, had previously seemed harmless. Such a casing could run the full depth of the well, anywhere from fifty to ninety feet deep. It was a neglected relic of a once useful source of water. Now decommissioned, there was no wellhead to manage the flow of water. There was no cap to protect it from debris. Only a makeshift cover—a flat rock—had been used.

The neighborhood in which Tanner Drive is located is known as the Permian Estates, and it is strewn with similar water wells. What was then known as the Texas Water Commission had begun keeping records on water wells in the area in the 1960s. Prior to that, state regulators did not require records on new water wells. While no record exists of the well at 3309 Tanner Drive, records from other wells at neighboring houses show depths reaching anywhere from seventy-five to ninety feet. Residents in the area paid for the extra expense of having contractors dig their residential wells in order to avoid the city's notoriously bad water quality. Midland by and large is known for the awful taste of its city water supply, and even into the 1980s, records indicate new water wells being contracted throughout the Permian Estates development.

Seventeen-year-old Cissy paced back and forth between the rusty brown well casing at the corner of the property and the house; the well casing was just six feet from a back fence that ran parallel to the alley, and it was about thirty feet from the house.

"Jessica!" she yelled out, sometimes standing above the pipe and at other moments leaning right into it. "Jessica!"

A faint whimper could be heard coming from deep beneath the ground, rising through the long, dank shaft. First responders later estimated that another fifty feet down was a pool of water.

Cissy ran inside the house to the telephone and dialed 911. Operator Lloyd Dunagan answered on the other end at the city's dispatch terminal. Panicked, Cissy gave few details, according Dunagan's recollection of the call; she frantically asked for help and gave directions to the address of the home.

"I just ran in the house to use the phone to call you," Cissy told Dunagan before running back to the uncapped well. "I got to get back."

Next door, at 3311 Tanner Drive, sixty-nine-year-old Maxine Sprague was working on a crossword puzzle, nursing the arthritic pain that had woken her a half hour earlier. The panicked calls for the little girl sounded oddly unlike the typical playful shouts of children. Maxine and her seventy-five-year-old husband, Raymond, usually enjoyed quiet mornings over a pot of coffee, with the cool West Texas morning breeze settling on their screened-in windows. It was normal for them to hear children playing next door at the home of Jamie Moore, Cissy's sister. Jamie took classes at Midland College during the day. Cissy filled in by babysitting when her sister was at school. All at once, Maxine heard the playful shouts turn into screams of terror as Cissy stretched her arm down the dirty brown pipe, reaching for her daughter.

"Oh, my gosh," said Maxine, springing from her seat as she heard the frantic cries for help from Cissy. "That's not the kids."

Maxine rushed into her backyard, still cloaked in her bathrobe. Her two dogs, Ish and Jojo, lapped at her feet as she ran to investigate. Peering over the fence, she could see a handful of children standing around the well casing; some were crying.

"Jessica fell in the well!" Cissy bawled.

Seeing the children, Maxine ran through a side gate, falling to her hands and knees beside the well casing.

"Jessica! Jessica, talk to me!" Maxine shouted. The three-year-old daughter of Lawana Keller stood by, looking at the woman screaming into the well. Another child, the ten-month-old son of Pete Starks, sat looking on without the slightest understanding of Jessica's newfound trouble.

Maxine heard what she could only interpret as a giggle as she put her ear to the well.

"I think she thought they were still playing," said Maxine later, surmising that the children only moments before had been playing near the well. Later on, others, too, would speculate as to how the children's play would lead to Jessica's falling down the well shaft.

Maxine waited by the well until she heard the sirens nearing. The chorus of sirens grew louder as other police vehicles joined whatever pursuit was taking place.

Patrick Crimmins, a twenty-eight-year-old reporter for the *Midland Reporter-Telegram*, had punched in around 5 a.m. that Wednesday. When Lamar students had heard the echoing sirens, Crimmins had just begun wrapping up his day. In West Texas, many newspapers operated as afternoon dailies. Each gray copy landed on doorsteps just before 5 p.m., as mothers and fathers steered their company-owned cars into their driveways. That meant that reporters like Crimmins needed to wrap up their news stories by 10 a.m. to allow copy editors to get their eyes on news text before placement. A few hours later, presses could roll out thousands of copies for that afternoon's edition.

The daily paper's news cycle required fast work for a budding reporter like Crimmins, who was still somewhat fresh out of Southern Methodist University (SMU) in Dallas, where he had not studied a lick of journalism. Crimmins left SMU with a bachelor's degree and headed for Lubbock for a shot at law school at Texas Tech University. It left a bad taste in his mouth, though, and he eventually threw in the towel.

"I got there, and I just hated it. I dropped out before the end of my first year," said Crimmins.

He left Lubbock for Huntsville, on the other side of the state. Though still in Texas, places in Utah are closer to Lubbock than that spot in the Piney Woods of Southeast Texas. There, he found an affinity for reporting thanks to the budding entrepreneurship of an angry gang of investors who gave Crimmins his first job to help fight the *Huntsville Item*.

In the end, the *Huntsville Morning News* lasted only one year, closing on August 1, 1984. Crimmins was tossed out along with the office furniture. It was a pride-swallowing experience—but then again, so was being a newspaper reporter.

In May 1986, Crimmins heard about an opening at the *Midland Reporter-Telegram* from his girlfriend who was working at the *West Texas Business Journal*—known as the *BJ*—whose offices were located in a loft above the pressroom of the *Reporter-Telegram*.

Wednesday, October 14, 1987, was a light morning for Crimmins. He hit the sheriff's office, located in the basement of the Midland County Courthouse, with a trimmed lawn and a miniature Statue of Liberty placed out front. The courthouse was the centerpiece of what was once a perfectly picturesque downtown. Crimmins then moved on to the police station. He made another check at the Central Fire Station, which was a straight shot from downtown along North Loraine Street. Nothing major to report.

Crimmins's *Reporter-Telegram* took on a role not unlike many newspapers in similarly sized communities. Much of the newspaper's editorial page coverage reflected the political voice of the largely conservative West Texans it served. The newspaper's competition until the mid-1980s had long been the *San Angelo Standard-Times*—a newspaper headquartered 110 miles east in West Texas's farm and ranch epicenter. Though Midland's demand for the San Angelo paper ebbed in the 1980s, reporters from rival papers still clamored to get to stories first. To reporters at the *Odessa American*, Crimmins's *Reporter-Telegram* was a "Chamber of Commerce" newspaper—a stinging insult to any real news person.

As Crimmins wrapped up and began heading out the door that Wednesday morning, the police scanners posted in the newsroom caught his attention.

Some remember a call for a girl stuck in a pipe. Others remember a call for a girl stuck in a well. Everyone remembers the first thought that flashed through their minds when they heard that a little girl had been stuck in a well. Most people pictured a rough rock-lined well shaft from a scene in a fairy tale, with a bucket lowered by a hand-crank. The reality was much different.

"3309 Tanner Drive, we've got a girl trapped in a pipe," a dispatcher called out for the closest police cruiser and EMS crew.

Rick Brown, the assistant city editor at the Midland newspaper, tapped on a keyboard at his desk. Life in Midland was not ideal for Brown. The chain-smoking editor belted out assignments as they were handed down as tips from higher management. He looked toward Crimmins who had already started making a move in response to the scanner's buzz.

Newspaper editors usually stick crime reporters with the emergency scanner, so it would have been difficult for Crimmins to have missed this call. His desk was just beneath a shelf where the scanner sent out its constant calls. It was his job to sort out the important ones—most of them were not so important as to necessitate follow-up. Alerts were sounded for ambulance runs on emergency calls for chest pains and minor crises. It was easy to develop a deaf ear to the scanner's constant hum of calls for EMS, fire personnel, and police.

The dispatcher grabbed Crimmins's attention along with that of fellow general assignments reporter Ramona Nye, who looked at Crimmins quizzically when the call came across. Crimmins returned the look, stuffing a notebook in his pocket and walking over to Brown's desk.

"I heard something funny on the scanner. I better check it out," Crimmins told Brown as started out the door.

Gary Ott, then the city editor of the newspaper—he would eventually become its editor-in-chief—looked on. Ott usually glanced up at reporters as soon as they came through the door. "Whaddaya got for me today?" Ott would ask each reporter upon his or her entry to the newsroom. For him, the inquiry was a playful social exchange, but for the reporters it signaled how he looked at the role of a local small daily reporter—like the engineer standing over the factory line waiting for finished product to roll out of the machines.

Ott's constant barrage and lack of tact could kill a reporter's will to live on a day-to-day basis. Crimmins summarized his own reaction to being cornered repeatedly by Ott for copy: "Fuckin' A, man. I just got here."

Unlike papers in larger markets, where a giant staff milks sources for stories to put on the back burner and slowly accumulates information for an eventual story, papers in smaller markets have plenty of space in the pages and a serious need to fill them with local news copy. At a metropolitan paper, reporters usually have to fight and play newsroom politics to get a story on a page; if they are really lucky, the hard work might pay off, and the story might make it on page one. In Midland, a reporter's byline on his or her first day of work could land on page one. In fact, the reporter could "own" the front page above the fold with a lead story and a sidebar story.

This training ground taught reporters to report quickly. And Ott, whose adroit editing often bailed young reporters out of trouble, had a power that even he did not likely recognize. He pounded the keys of his computer keyboard as though they were still the mechanized hammers of a typewriter. The banging raised eyebrows from the newsroom staff, but more often than not he was too narrowly focused on the editing task at hand to notice the attention drawn by his typing racket. Ott's better-known published work came in the form of a local opinion column—more of a collection of rambling thoughts—that harbored a wit and skill that made him memorable, even to an eventual president who picked him out of a crowd while campaigning in Ohio in 2000 to say hello and acknowledge him by name.

Steering his way out of the newsroom that morning, Crimmins did not have much time to get his copy back by 10 a.m., to make the afternoon edition. If the incident warranted coverage, he would have to hustle. Crimmins's trip to Tanner Drive was a quick one. As Crimmins arrived on the scene, the alley running along the rear of the little home where rescuers gathered was spilling over with emergency vehicles. Like any other alley in Midland, the dusty dirt paths host city trash containers, litter, and vicious red fire ants that roam like crawling tiny swarms.

In 1987, the *Midland Reporter-Telegram*'s offices were located at the eastern edge of the city's downtown. The handful of skyscraping buildings lent the city its nickname the "Tall City," and the skyscrapers stood out against the desert landscape that was otherwise as flat as a tabletop. Since the early 1950s when the large corporate oil interests established many of their headquarters for their Permian Basin oil operations in downtown Midland, the small town had become home to an oddly corporate appearance, a portrait of irony, given the backdrop of a rough arid terrain. The corporate skyline is visible from more than twenty miles away today as the flat, open plain gives way to the view.

In 1987, the flight from Midland's downtown was underway for many of the corporate oil interests that had ceased operations in the Permian Basin. The previous year in the oil business had been a tough one—although just five years before, the town had basked in the glow of economic comfort afforded by a historic oil boom.

The oil boom times in Midland were robust until the mid-1980s, and the economic impact had resonated throughout the region. The residual effects from the boom multiplied beyond West Texas and spread success across Texas. This economic fortune made itself especially clear in the state's budget, where waves of new tax revenues fueled the state's growth.

At the time, 23 percent of the state's tax revenue was generated from the black muck flowing in striations 10,000 feet below the earth's surface.

Since the early 1900s, this tax base made Midland and the surrounding oil-producing counties powerful, attracting political fundraisers for hopeful new candidates as well as for established incumbents. Midlanders knew it, too. They had grown accustomed to it, and it felt to many as though a special providence existed in this place that did not exist in other towns like it. It is a strange notion, given the terrain, the city's size, and its people. It was a long time coming. The power of Midland as the economic epicenter of the West Texas Permian Basin grew immensely in the early twentieth century, attracting business investors from around the world, but unlike the more widely known great economic cities like Chicago, New York, Houston, and Dallas. Midland's wealth per capita constantly attracted news writers who frontloaded stories with descriptions of Midland's high-dollar luxury car dealers in flowery narratives for national publications. Such rapidly accumulated wealth attracts storylines depicting subjects with an extravagance that comes across as satire were it not for its truth. The subjects of these stories could hardly be blamed. The speed with which an entire economy formed—from the broader worldwide oil business to the more finite oil-producing regions—was perhaps faster and more robust than any in history.

In 1900, Texas produced less than a million barrels of oil annually—1.3 percent of the total US production. On January 10, 1901, Texas began its climb to its position in the world as an energy behemoth when the Lucas No. 1 Spindletop well blew oil about 100 feet into blue Beaumont sky. Initially, it was estimated to be producing 75,000 barrels of oil daily, which would have been 39.4 percent of the average daily US production. In 1902, the well would produce 17.2 million barrels for the year. Alone it produced 19.7 percent of the total US production for that year. It would produce oil until it was plugged in 1990.

While the economy of the nineteenth century had been one dominated by railroad barons and the thrust of westward expansion, Texas would set itself as the oil and gas foothold for the American economy like no other state. In 1900, the state's GDP, dominated mostly by cattle and cotton, sat at $119 million. By 1940, the state's GDP had skyrocketed to $29 billion. The 240-fold increase compared favorably to the overall US GDP's 24-fold increase over the same period.

The West Texas oil field growth was not dissimilar to other industrial booms. The pace of growth could be compared with that of many other cities that experienced a sudden commodity discovery. Attracting an influx of investors, workers, and a service industry, West Texas beckoned legitimate prospectors and bankers as well as enterprising upstarts. In the Big Lake field, southeast of Midland, where the Santa Rita No. 1 would strike oil in 1923, prospects were so optimistic that seventeen additional wells were started before the first well would see its own full year of production.

Populations in small towns like Big Lake exploded. Between 1920 and 1922, Breckenridge, Texas, grew from 1,500 to 30,000 people. Odessa grew from 750 to 5,000 between 1925 and 1929. And Midland's thrust at midcentury was unstoppable as new discoveries in the Permian Basin lured oilmen and industry to establish their West Texas headquarters in the Tall City's high-rises. In 1950, with the state's annual production at 800 million barrels, Midland County alone produced about seventeen million of those barrels of oil. Worldwide, producers pumped 3.6 billion barrels per year, with the United States contributing a large percentage of that international production, and Midland producing a significant share of that overall inventory. These production levels do not include the regional wells, which make up the rest of the Permian Basin north to Hockley County, west to Howard County, and south to the border with Mexico. In 1950, while the United States accounted for 50 percent of the world's oil production, Texas alone contributed up 22 percent of the world's overall oil production. And the Permian Basin accounted for 11 percent of all international production. Midland County, one of the lesser oil-producing counties of the Permian Basin, accounted for 0.47 percent of the entire world's oil production.

The Permian Basin reached its production peak in 1973, topping out at 56 million barrels of oil per year, or about 1.8 million barrels daily. But by March 1983, the price for a barrel of oil sputtered, falling from $34 to $29 and stoking concerns that economic trouble might be on the way. Ups and downs were to be expected. For the entire twentieth century, Midland and the broader West Texas economy had endured the roller coaster ride that was the Texas oil business. Support for an economy that existed in oil's halo flourished. Men who had struggled to graduate from high school ran their own successful companies in the service of the oil economy. They owned small dump truck or backhoe companies. They started niche construction operations. Others built oil companies that supported drillers in operations like safety and security. In areas where oil sold at high prices, these companies flourished at the margins.

The Permian Basin has always been stitched together in a patchwork of small and medium-size companies serving the oil and gas industry. They are not always the big corporate giants like Exxon, Chevron, and British Petroleum. Many are smaller equipment outfits with one specialization or another. Other support companies provide tools and services as ancillary as protective clothing, helmets, and gloves or as aligned as pipes, chemicals, drill bits, or the oil rigs themselves. Some of these companies consist of former roughnecks who scraped together enough savings and business contacts to start their own small businesses—trucking and hauling operations with specialization in managing certain materials like gravel, for instance. Others are engineering outfits with small teams generating small fortunes.

When oil booms, the economy's bustle is felt across the region, and small companies emerge with noticeable profitability. Others realize innovations and solutions customized to suit large oil operations, and with expanded margins reducing the risk, this opens up opportunities for entrepreneurial minds. Some of these emerging services are premium priced or sit at the fringes of the workflow around the energy industry. The Permian Basin thrives when these businesses proliferate. But their utilization as a frill shrouds their weaknesses—until oil prices drop.

Longtime West Texas oil field veterans like Morris Burns carved out such solutions in the 1970s and 1980s boom. Inspired by an entrepreneurial thrust, Burns took over a small Midland company, FerreTronics, in the late 1970s, eventually becoming its president. For FerreTronics, Burns sold and serviced a gas detector that would alert rig operators when the mud coming out of a drilling hole held hydrocarbons. From 1979 to 1982, Burns worked as president of the company and saw all twenty of its detectors in the field. The twenty units ran for $100 per day, and Burns's salesmen in the field took 50 cents for each mile they drove to and from the site to install, replace, repair, or remove the unmanned mud-logging units.

Pulling in two thousand dollars a day on a patented product in a booming market meant a steady flow of income for Burns and his company. No fewer than ten of the detectors were running at any one time during the boom. In Texas alone, 1,000 rigs were up and running, ensuring a wide-open market for FerreTronics equipment and future growth.

The Texas oil economy is popularized by programs depicting big oilmen making million-dollar deals. While those types do exist, natives to the community recognize them mostly as caricatures. The more accurate representation, however, features a bustling region with men and women more like Burns—individuals putting together small contracts, living on the fringes of a billion-dollar industry, running their own companies that together support thousands of full-time employees.

"We were blowing and going there for a while," Burns said.

Signs of a deflating boom were slowly showing through, in the early 1980s, and executives like Burns began to take notice. When his clients started cutting back on expenses in response to falling oil prices, Burns knew time was short for his company. FerreTronics' devices existed in that thick margin between boom times, when caution fell to the wayside, and lean bust times, when drillers focused on only those inputs that were necessary to suck crude from the land.

As Burns's company faltered, oil field pull-back started to affect other companies as well. It would take more than a couple small oil field service companies for the broader economy—beyond commodities traders tuned to the daily oil price—to take notice, however. Like many economic bubbles, a large giant would have to fall and make a noise loud enough for everyone to hear—and that is what happened when First National Bank of Midland began cracking.

West Texas banks were loaning 100 percent of the money it took to buy a drilling rig plus $100,000 toward the first year of operating the rig. In 1981, there were 4,100 rigs running nationwide. In lean times, when the price per barrel of oil was moderate, that number would fall to about 1,250.

It was not easy to put a rig together for exploratory drilling. It took well over a million dollars to get an entire operation up and running. Even finding the equipment to compose a whole oil or gas exploration operation became difficult. Rigs had been sold for scrap or were cannibalized to make one good one. Many of the parts had been sold to overseas producers who were just learning how to pump oil from the arid stretches of their own deserts. In 1981, Saudi Arabia produced less than two million barrels of oil a day. The overabundance of supply still failed to outpace demand, and commodity traders purchased oil for a price at which investment in oil exploration could still be feasible. There was enough room in international demand for American producers to benefit lavishly from domestic production. It would not last much longer.

From 1982 to 1984, oil field service industry vendors like Burns started to feel a pinch from falling oil prices, though oil prices remained moderately steady through mid-1983. By 1984, Burns had scaled back his FerreTronics operation to himself and two other employees. By 1986, all that was left of FerreTronics was Burns, a Jeep Cherokee, a pocket pager, and a mobile phone. The first to go during an oil bust are the budget line items such as services that make production more comfortable or efficient but that nonetheless can be done without. These companies exist in the service economy as a part of the oil and gas ecosystem; they benefit from big money multiplying to vertical operations when times are good, but they are the first to suffer when oil prices drop. Operators scale back their own employees to bare-bones personnel, putting together teams that get the job done with the least possible expense.

The scaling back of the service industry workforce meant that many of those once working in the oil business in Midland were no longer employed in the oil business at all. They began to submit claims for unemployment benefits in droves. Although some companies were successful at staying in the business, those struggling were lower- to middle-class workers—some of them roughnecks who had worked the field and some of them white-collar geologists.

Still, the clock was ticking for banks as oil producers carried around debt loads with no way to pay off their loans. Producers who relied on credit to make their operations run were up against a wall, and many of the families who tried to make it work scraped money out of savings while hoping for an economic turnaround. Others took jobs well outside their experience.

At the time, a morbid joke circulated that illustrates the depths of despair for many professionals serving the oil economy. "How do you call a geologist in Midland?" the joke begins. "Oh, waiter!"

Writers covering West Texas boom town banking mentality attributed the borrowing spree to a sort of ancestral hubris. In 1975, famed Texas author Larry L. King described the tendency to lend liberally to these Texas landowners and their descendants as a process that did not necessarily go on unmeasured, although practices certainly failed to rise to high standards. "They [banks] did not urge money on you just because you wanted to hunt oil and thought you knew where to find it. But if you could show good prospects—a good geological survey, proximity to earlier drilling successes, an option to explore whether signs look good—then Midland banks would finance you to their long-term profit," King wrote. King, who was born and raised in Putnam, Texas, a few counties east of the Permian Basin, wrote extensively on West Texas. He had been a reporter in Midland and Odessa before moving to Washington, DC, to serve as an aide to several elected Texas officials. King was venerated as a sort of satirist and Texas expat and was celebrated for his critical, yet loyal, views on the Lone Star State.

Other journalists had a more pragmatic view of Midland and its boom-and-bust-prone population. Skip Hollandsworth, the longtime *Texas Monthly* columnist, covered many West Texas oilmen, particularly in its latest boom of the 2000s, which started coming back to life around 2006 to 2011 before leveling off on a high for several years.

"They talk about how they hate the busts, but deep down, the Midland oil crowd loves its life. They love surviving the downturns. They love knowing that they nearly lost everything and yet still rose again. That's why they're so fascinating," Hollandsworth wrote. He went on to describe an afternoon he spent following Midland oilman David Arrington. Known for his acquisition of a vast collection of 650 Ansel Adams portraits, Arrington is the poster child for eccentric and loud Texas oilmen. After graduating from Texas Tech University in Lubbock with only two classes that covered oil and gas, he arrived in the field at the bottom of the industry's low point. Starting as a land man at a small company, he spent much of his spare time researching prospective properties he could wildcat. In 1985, he struck oil near the small town of Kermit, south of Midland. Eventually he sold off his Permian Basin properties, pocketing $25 million. He would then go on to invest in the booming Barnett Shale area near Fort Worth. He eventually sold part of that production to Chesapeake Energy for $209 million. He sold the remaining part to XTO for $450 million.

Hollandsworth watched Arrington signing royalty checks—payments to landowners or partial landowners with mineral rights—or to other investors who had invested in Arrington's ventures.

"The stack of checks was about five inches high," Hollandsworth said. Arrington let Hollandsworth get a glance. One check was for more than a million dollars. Several were worth a few hundred thousand dollars. And then another one for a million or more. "And that was just the checks for one month. The next month, he did it all again. Someone out there gets a check every month simply because he was lucky enough or smart enough to invest in an Arrington project."

For those who look upon an afternoon of signing a massive stack of royalty checks as a practical project for a spoiled, rich, and lucky guy—doling out millions of dollars in what seems to be a whimsical task as blasé as returning calls—perhaps they have a more cynical eye than others do. Or perhaps they have a more straightforward grasp of reality and see that the complex puzzle pieces of the oil game fit together quite simply. Either way, as Hollandsworth described it, this place made Texas what it is today and what it will likely be for a long time to come.

As Hollandsworth put it, "The oilman built modern Texas. They gave us our character. No matter what else we do, no matter where else we go, there will never be anyone like our oilmen." When Saudi oil interests began to insert themselves into the picture, however, there would be fewer dreams of joining that enviable class of big time Texas oilmen.

In 1986, King Fahd of Saudi Arabia grew tired of OPEC's (Organization of the Petroleum Exporting Countries) limit of eighteen million barrels of oil production per day. The head of Saudi Arabia's oil ministry, Sheik Ahmed Zaki Yamani, became Fahd's target, since Yamani had supported the production levels and even urged countries belonging to OPEC to limit production. Yamani was a thorn in the side of Saudi landowners with vast mineral reserves—despite his having held his OPEC position since 1962, and despite his having helped mastermind the development of OPEC. Saudi leaders looked past the fact that Yamani had effectively raised the price of oil from where it was in 1972—around $2 to $3 a barrel—to its drastic rise, starting in 1979, to around $13 per barrel. By 1981, oil would reach $39 per barrel, aided partly by decreased production due to the war between Iraq and Iran.

Yamani's fall resulted in Saudi Arabia increasing its production. What resulted was a flooding of the market. Demand stayed steady. Supply increased. And the price of oil dropped like an anvil from the sky, crashing hard on Midland.

From 1981 to 1986, half a million American jobs slipped away from the oil business. The rig count in Texas plummeted from almost 700 to 311.

Many of those rigs lost were in Midland.

In West Texas, the rumblings among those who depended on the banking system were growing more troublesome. Still, the rumble did not first start its quake in Texas but, instead, farther north in Oklahoma.

On July 19, 1982, Penn Square Bank in Oklahoma City announced its failure. Considered one of the most aggressive in the oil and gas industry, it was also home to a wild chief lending officer whose outlandish behavior served to underscore his bizarre risk-taking deals. Other banks that had heavily lent to oil and gas ventures soon began showing signs of trouble. On August 6, 1982, Abilene National Bank failed; it was ultimately forced to merge with Mercantile Texas Corporation.

Just a week before, the head of First National Bank of Midland secured a $100 million loan from the Federal Deposit Insurance Corporation (FDIC) in anticipation of finding a buyer, but the weight of its debts had grown too heavy. With 76,400 accounts, losses from defunct energy loans reduced reserves at the bank to $862,000 by the end of September 1983. A year earlier, the bank celebrated reserves of $122 million.

In the same year, delinquent loans doubled to reach $31.4 million. Soon, a cadre of the FDIC bankers took over, setting up shop to manage the bank's $1.3 billion in assets.

CHAPTER 3

THE ARRIVAL

AROUND 9:30 A.M. ON WEDNESDAY, OCTOBER 14, 1987, first responders began arriving on the scene at 3309 Tanner Drive, where Jessica's faint whispers echoed up the long well shaft. Pulling up to the house, they rushed to the front door, dashing through the bits of Bermuda grass, now brittle and crispy with the cooling fall weather. Cissy stood at the door frantically waiting for their arrival.

"She's here in back!" she yelled out, salty streaks of tears streaming down her cheeks. Cissy led the group to the well, a rusted brown pipe in the backyard. She stood over it, pointing toward her daughter. "She fell down right here!"

Midland police officer Bobbie Jo "B.J." Hall arrived in a white patrol car, sirens blaring. The thirty-two-year-old Hall had just hustled to the backyard only to confirm that the call to 911 was indeed not a hoax or strange exaggeration. The dispatcher's initial call for officers had sent shivers through Hall when she had heard the description only moments earlier. Seeing the small rusty pipe, she realized that it somehow held within its depths a child. Her heart pushed into her throat. She rushed to the well pipe, fell to her hands and knees, and reached in up to her shoulder. Perhaps her reach would reveal the girl lodged near the top of the shaft. Nothing. No sign of a child's grasp or even a wisp of hair. Hall, placing her ear to the well's opening, listened for a moment; she could hear the faint sound of a girl's soft voice.

Around the same time, paramedics Tim Owens and Bill Walker maneuvered an ambulance down Tanner Drive. They steered their vehicle into the back alley, dipping right, then left, into each tire track's sunken path. The tires crushed the hardened edges of the ruts, breaking

CHAPTER 3

the ribbons of dried mud into refined bits of dust, curling them up behind the ambulance in a wafting cloud. The last rain had done little to settle the landscape after the blistering summer, and the cool of the fall had yet to fully embrace West Texas.

A few miles away, Sergeant Andy Glasscock of the Midland Police Department (MPD) made his way out of a training session on new emergency response protocols when he heard the emergency call about a girl in a pipe coming across the airwaves. Glasscock, wearing black pants and a button-down shirt with a dark tie, walked to his patrol vehicle, a beige unmarked sedan parked near the downtown police headquarters on Texas Street.

Although he was a sergeant and a detective, Glasscock's position did not make him part of an elite investigating force. Since 17 percent of the MPD was made up of detectives, few gave the status any credit. Still, he was sure on his feet and secure in his role as a ranking police officer on the force who knew his way around the city and its police culture. He had graduated with a degree in criminal justice from Angelo State University, located 100 miles away in San Angelo and renowned for its law enforcement program. While cocksure and a little boyish, with an immature sense of humor, Glasscock maintained a reasonably good reputation. Heavyset with a classic cop-cropped mustache, he looked the part.

The call had caught Glasscock's attention, piquing his curiosity. Nothing about the call for Jessica necessitated his presence. "I got in my car and just decided to go there," Glasscock recalled later with a shrug, unable to pinpoint exactly why he went to the scene. Outside of the off chance that he may be able to help, he felt a general curiosity following the call for a rescue of "someone trapped." That is all he remembers from the original call. It accords with the memories other first responders reported who heard the alert go out on the radio that day.

En route, Glasscock twisted the volume knob to his cruiser's radio and heard Officer Hall call back to dispatch.

"Yeah, there's somebody down here," she said, confirming a frightening prospect but leaving an ambiguous mental picture. Few really had seen a water well on a residential property. A windmill standing guard over a wellhead here and there on a small plot of acreage would certainly be normal. But a well in a residential backyard seemed odd and inspired a vision of a wishing well from a children's fairy tale.

As Glasscock wheeled around the corner onto Tanner Drive, several MPD cruisers already clogged the street. Not the first to respond, but likely the first of higher rank than others on the scene, he walked right up to the well casing and tilted his flashlight into the pipe. He gave a few nods to those in the group trying to assess the situation. Huddled around the well, some on their knees and trying hard to listen and get a bearing on the little girl's exact condition and position, several turned and looked at Glasscock as he made himself known. Glasscock considered the small well casing, not more than eight inches wide; he was perplexed by the improbability of any child falling into it.

"Give me a break," he said to himself and then looked to Hall and the others.

"Nobody can be down that pipe," Glasscock snorted with doubt. "It's not big enough."

His retort met little response, but the officers and paramedics at the scene could not argue. Cissy stood by, assuring them that her baby had slipped down the shaft when she turned her back for a brief few moments. What else could explain the missing baby's whereabouts? She certainly did not waltz her way out of the backyard unassisted.

"When I left them, they were playing on the porch with some of the reeds and stuff," Cissy explained, noting the yucca that had been shedding and had captured the kids' attention as she had gone to answer the phone and use the bathroom. The large plants grew in a grouping of other loosely kempt vegetation that had grown taller than normal along the fence. "When I came back, I counted the kids, and they were all over by the hole and mine was gone. I thought maybe she was playing on the other side of the fence or behind the reeds, so I went and looked."

The crew passed blank looks back and forth to each other while she explained. Then Cissy, at a loss and not knowing what else to do, moved off and sat down with her chin in her palm while a handful of children played with toys in the yard, and the first responders swung into action.

CHAPTER 3

At that very instant, *Reporter-Telegram* photographer Curt Wilcott snapped a photo of her before rushing back to the newspaper's offices.

Glasscock bent down on a knee, peered down into darkness expecting to see a sign of a child wedged perhaps a few feet down. Using a flashlight, Glasscock and Walker could see what appeared to be reeds, but no Jessica. The notion of being that far down in the hole—and possibly blanketed with a swath of vegetation—led Walker to concerns about suffocation. He found the nearest water hose and began developing a plan in case Jessica could not breathe fresh air.

"What's the child's name?" Glasscock called out to Cissy, who was still sitting across the yard.

"It's Jessica," she called back, getting up and walking back to Glasscock and the others.

Glasscock turned back toward the well casing.

"Jessica!" he called out. "Jessica!"

His voice boomed, dissipating into waves down the shaft, bouncing along the rusty walls. His call was met with the faint, gentle moan of a whimpering young girl. It was the first instance of contact between Glasscock and the child, setting in concrete the reality of what he was facing.

"My God, there's a child in there," he thought to himself; and thus began the fifty-eight-hour stretch at the well that would eventually thrust him into the spotlight as one of the key players in the effort to save Jessica McClure. His stomach turned, and his heart began to race, the child's echoing whimper climbing its way up the pipe. Glasscock would later say that in that moment, the reality of a child's life hanging in the balance had hit him in a place where emotion normally did not reach while on the job. With a young child of his own at home, Glasscock said the girl in the well was no longer some anonymous youngster, and he emotionally connected himself to the unfolding crisis as a father.

Walker, one of the first paramedics on the scene, walked over to the well and slipped the end of a long green garden hose down the shaft. Snipping the metal ends off, he wrapped duct tape around the remaining end, affixing it to an oxygen tank, which he then opened.

No one counted the rolls of duct tape used to facilitate the rescue by the day's end, but Walker's move was the first of a number of resourceful measures taken in answer to predicaments that threatened to kill Jessica; the answers would continue coming as rescuers ran into problem after problem.

Now that men and women of law enforcement and EMS were fully on the scene, they were able to wrap their heads around the challenge they faced. The reports into dispatch from the first responders started to open some eyes after what was almost dismissively described as a call for a girl trapped in a pipe. It turned into a true rescue call for a girl trapped twenty feet down a dark and forbidding well shaft.

More rescuers would be needed.

Dave Felice and Chip McCoul sat in an incident command class at Central Fire Station near downtown Midland at 9:30 a.m., a couple miles away from their assigned MFD station—Station 2. The idea of having an incident command post to establish coordination at a scene was fairly new among first responder communities and their leadership. Rarely would a coordinated chain of command at the scene of an incident be necessary, but as the West Texas population continued to grow, so did its oil and gas infrastructure. And so, too, did its need to maintain a well-trained and sophisticated emergency management program in the event of a large-scale emergency.

Felice and McCoul heard the dispatcher when the call about Jessica McClure drifted across the airwaves. The run on the rescue would be assigned to other crews at Station 6, a small post at the southwest corner of Midland Drive and Thomason Drive. As the firefighters and paramedics sat in the classroom and waited for dispatch to call out which station would be the one to go to the scene—all of them pausing and looking up toward the speaker in the ceiling—a few thought it might simply be salvation from Bob White's incident command class. But the relevance of White's incident command training and prioritization would become apparent in the next few hours.

CHAPTER 3

After five minutes more of the class, White went down the hall to dispatch to figure out just what the call about the girl in the pipe could have been about—some of the paramedics and captains, like Felice, caught a few of the dispatches from the scene on radios they had with them in the classroom.

"All right, guys," White announced, coming back into the room. "Go back to your stations. We're done for the day. We'll pick up class later."

Felice and McCoul shook off the early cutoff and climbed on board their station's main engine to make a stop for the station's daily groceries before heading back to Station 2. McCoul was behind the wheel, and Felice sat shotgun. They headed to M System—a chain grocery store on Andrews Highway.

Dave Felice fit the stereotypical image of an American firefighter—thin and taut, with a dark and neatly trimmed mustache. His frame supported a layer of muscle well defined after training for marathons and firefighter physical fitness tests.

In the M System parking lot, Fire Marshal Jerry Petree pulled up in his cherry red sedan as Felice and McCoul returned with baskets of brown paper sacks stuffed with sustenance for the twenty-four-hour shift.

"Dave, you're going out to the scene," Petree called as though the developing rescue had already turned into an event the entire fire department already knew about. "Get in."

Working at Station 2 was no walk in the park. There were plenty of calls for runs along the busy US Interstate 20 that undercuts the southern edge of Midland's industrial district. The strip of highway—its westernmost section covered by the Odessa Fire Department—facilitated the travel of thousands of tanker trucks and oil field service units traversing the desert edge of the Permian Basin. The constant attention to the safety risk of the vehicles meant that Station 2 was always on the ball.

"What? You mean out to Tanner Drive?" Felice asked, wrinkling his eyebrows, with a tuft of mustache rising and falling with his words. They had heard the rescue call get assigned to Station 6 before leaving the training at the central station.

THE ARRIVAL

Petree replied, bobbing his shoulders, "Yeah, they want you out there."

Felice waved off McCoul, sending him back to Station 2, where other firefighters like Robert O'Donnell waited, planning to finish out their shifts like the other nonresponding firefighters in Midland. After all, the whole town still operated no matter which girl fell into what well.

Glasscock was already at the scene when Patrick Crimmins arrived, hurrying to get as much information as possible back to the paper for a deadline that had passed fifteen minutes earlier.

Though Crimmins worked just a few miles from the scene, a news crew from KMID-TV Big 2 News pulled in first. Cameraman Phil Huber was already shooting footage when Glasscock and the rescuers thought about needing to hear Jessica more clearly. The two watched as rescuers began formulating a plan, calling for this equipment and that equipment before pulling down a clothesline. Along the fence separating Jamie Moore's backyard from Maxine's, a growth of bamboo shoots sprang oddly from the ground, serving as the source of the long, leafy reeds. Huber focused his lens on the various aspects of the backyard as well as on the developing rescue. What would go in the first package of stories could be anything, and getting as much of the environment as possible on tape became priority one as responders looked down the well toward Jessica.

In West Texas, spot news drives viewership. With the *Reporter-Telegram* operating as a daily newspaper, a broadcaster's effort to compete with the print product was still driven by scoops rather than by details. Television reporters dismissed print stories on the same subject no matter the level of detail curated in the print product. In the West Texas television market, cameramen like Huber were the best bet for working scoops in of the field, putting their watchful lens on the story and following up with sources and comments later.

No reporter in the local news market would ever cover a story of this magnitude again, while even fewer in regional and national markets would ever meet a story with a similar significance until a

37

series of passenger jets would leave their scheduled flight paths in September 2001.

"Do you want a scoop on this, or do you want to sit up there?" Glasscock called out to Huber, who sat his camera on the edge of the four-and-a-half-foot fence of Maxine's backyard.

Huber gave Glasscock a look, trying to figure out what the cop was talking about.

"We need a microphone down in that hole to hear that girl," Glasscock explained to Huber, pointing a stout finger into the darkness of the shaft.

The complexities of journalism present many rules for those covering an event. Huber's facility in responding to Glasscock's request meant assisting in saving a little girl. It also meant he would become the story—or at least part of it.

"I don't know if I can do that," Huber replied.

Glasscock knew the media in Midland well. He called reporters when a story started cooking at the department, and he knew when it would make a juicy piece for reporters. If reporters treated him well, he treated them well. Such action might be as simple as giving a call in advance to a reporter the next time a story came around.

"Look at it this way," Glasscock explained. "You're gonna get exclusive footage."

The faint whimpers that had wafted up the shaft before now came to the top in full force, amplified by the acoustics from twenty feet of pipe and Huber's microphone. With the clarity of her voice, rescuers determined that Jessica was in fairly good condition. Glasscock would later describe the torment he felt hearing the sweetness of her voice coming up through the darkness.

"She sounded scared, but healthy," said Glasscock.

With sounds made so clear, her vulnerability became even more real, and it also became apparent she could shift in the well at any moment. She would be lost. The rescuers knew they had to act fast to dig her out. But besides not knowing how far down she was, officers could not fully determine just what kept her from slipping farther.

"Let's get a backhoe in here," someone yelled.

THE ARRIVAL

Obtaining a backhoe tractor was the easy part. However, the heavy equipment with its long arms with three joints across it forming multiple elbows to cant forward and backward requires a skilled operator. Getting into a backyard—with fencing and overhead power lines—is another challenge altogether.

Bill Bentley, a stout cable technician for Dimension Cable, drove a company truck into Tanner Drive's alley after Dimension's dispatcher called, preparing to snip cable lines to give the backhoe more headway. It was Bentley's job to utilize his technical prowess to deliver cable services to area neighborhoods. Scaling poles and hoisting himself high in the air in a bucket on the end of a dual joint arm from a truck, Bentley connected hundreds of homes in the area. Working at such heights and squeezing his way between tree limbs diminished anything resembling those traditional fears other human might possess related to heights.

"The police department said we need a tech out at 3309 Tanner," the dispatcher had ordered out over a CB radio to Bentley.

"What for?" Bentley asked, a note of apprehension slipping into his words with the mention of police prompting the call.

"I don't know. Just meet 'em," the dispatcher said.

Bentley, a West Texas native, had become an avid caver in his teenage years, worming his way through narrow passages hundreds of yards beneath the earth's surface. In a territory not known for its subterranean terrain, Bentley's skill and experience in caving came from his travels, which took him all over the country to explore caverns. A hobby to some, caving was a keen pursuit for Bentley, who documented his trips underground with other small groups. The washed-out photos of his underground crews with their gear—harnesses, helmets with lights and long ropes—are now scattered across the internet. The caving addiction Bentley cultivated meant he had an array of equipment like hard helmets with mounted headlamps as well as rappelling harnesses, carabineers, and ropes in his possession.

CHAPTER 4

LIFE UNRAVELED

Moments after the Oklahoma City Bombing

THE LIGHT FROM THE LIVING ROOM TELEVISION FLICKered with images in front of Robert O'Donnell as he sat in his mother and stepfather's home in a rural area outside of Big Spring, Texas, mesmerized by television reports coming in a constant stream from Oklahoma City. The words he heard triggered his concern, and he focused on the impending needs of those whom he saw working in the wake of the bombing that had just ravaged the community and the entire nation. O'Donnell watched the television news coverage—he saw the same attention and praise that had once been lavished upon him, championing him as a hero, now being lavished on the first responders. A full weekend of the news coverage continued, and by Sunday night, April 23, the constant stream of updates became overwhelming. These reports, coming from so far away, seemed to sit right at his doorstep. So familiar. So real. But a state and a lifetime distant from him now.

The 10 p.m. news in Midland combines coverage from neighboring Odessa, which helps only to fuel the feud between the cities that are now of fairly equal sizes. The cities are separated by twenty miles—not nearly as significant as the years of difference the respective citizenry feels about the other. O'Donnell watched as the weekend's developments spilled out from the screen, splashing onto the floor, waiting for him to lap it up.

This sort of wall-to-wall news coverage with camera crews making their way onto a disaster scene was familiar to O'Donnell. With cameras and reporters orbiting, footage of the Oklahoma City bomb site was only hours old. O'Donnell had been around for the birth of this kind of news coverage. This kind of reporting had yet to exist in full when he and hundreds of volunteers swarmed the Jessica McClure rescue site a little more than seven years prior. Now, though, Oklahoma City was the new center of the universe, its new breed of heroes captured in ten-second sound bites sent across the world as tasty little morsels. O'Donnell was on the outside of that orbit. And he knew it.

Yvonne, O'Donnell's mother, could hear the television news as she finished up a few Sunday night chores around the house, the foremost being a load of laundry for her son to haul with him back to Lubbock for the coming week. He had been on the job only a few days, but he had a new pickup, and she was helping him make the final preparations for his trip.

"Where are the shotgun shells?" O'Donnell called out from his bedroom, preparing to scour other parts of the house.

"For what?" Yvonne yelled back, folding her son's clothes neatly and then piling them in the living room.

O'Donnell was due in Lubbock early in the morning, and Yvonne still had some of his clothes in the dryer, waiting for her to fold and pack.

"I hear a rattlesnake outside," O'Donnell said.

He and his stepfather poked around the front yard for a while looking for the snake, a beam of light shining from a flashlight held high, looking for something that may or may not have been there. The two gave up the search and O'Donnell wandered in again a few minutes later.

Still, O'Donnell did not give up his search for the shells—and the gun.

"Where'd that shotgun go?" O'Donnell asked.

Yvonne would later explain to a reporter for *New York Times Magazine* how that evening unfolded. As he probed closets and corners of the home, looking for the gun, she repeatedly told her son to leave it alone. She told O'Donnell, "It's OK where it's at. There's no snake."

Yvonne went to her bedroom, not thinking about the issue any longer. O'Donnell appeared to have given up looking for the shotgun.

Just what O'Donnell saw on TV that spurred his decisiveness is unknown, to be sure; but Yvonne was certain her son had simply gone to bed. She thought he had given up on the search for the gun and had resigned himself to face another day with the hope of making this new start in Lubbock.

Getting to this point, though, had been a circuitous path. He had tried to make a clean beginning. But like many things in life that commence to crumble, cracks in the foundation had begun to form despite efforts to make critical repairs. Before he resorted to living with his parents on the family ranch, O'Donnell had attempted to start over on the other side of the state, in Huntsville—about as far as you could get from Midland without leaving Texas—working as an ambulance driver for the Texas Department of Criminal Justice's prison system, in the heart of death row. It was a purposeful and decisive move.

On July 12, 1993, O'Donnell had taken on a full-time position that kept him somewhat closer to his former role with the MFD. In Huntsville, he carted prisoners back and forth for various medical treatment. O'Donnell was in a faraway place where few remembered the heroics of the Jessica McClure rescue.

His job search had been arduous. He had sent résumés to fire departments all over the state, which had resulted in little, if any, attention. His four-page résumé was a detailed accounting of his life's accomplishments, and one would have assumed its contents would reveal the experience of a seasoned veteran. Finally, the Texas Department of Corrections responded, offering him $1,721 per month to work for the prison system at the Ellis Unit in Huntsville. It was the closest O'Donnell could get to a fire house since getting fired from the MFD in the early 1990s, after a collection of reprimands for showing up wobbly and dosed up on prescription drugs.

He had thought things could be different in Huntsville. But they were not.

According to his brother, O'Donnell considered committing suicide on Christmas Eve 1993. Instead, O'Donnell waltzed into his boss's office and dropped a hastily scribbled note down on his desk. "I hereby tender my resignation affective [*sic*] 12-24-93 at 1500 hours," the note read. It closed with his signature, which had three loops at the end of his name where there should have been just two.

His brother, Ricky, soon after took a call from him.

Robert managed to muster the words to ask for help, although a handful of pills already sat in the pit of his stomach.

It did not help that his job at the Ellis Unit had not been going well. Prior to his Christmas Eve resignation, O'Donnell received several unsatisfactory marks for responsibility and dependability.

"Nobody really knew what he had done [with the Jessica rescue] down there [in Huntsville]. And they didn't care. And he didn't like just being a regular person again," said Ricky.

O'Donnell's performance report from a superior noted his inability to adjust to the new setting; clearly, there was no extra slack given for his past notoriety and performance in Midland.

"Mr. O'Donnell never seemed to be able to adjust to the provision of emergency care in a correctional setting. He was not able to get to work on a consistent basis, and he had trouble accepting responsibility for his actions," wrote Greg Dickens in his evaluation of O'Donnell.

To a certain extent, O'Donnell had improved things for himself while in Huntsville. Considering the degree of his addiction, his consumption of pain medication had not been a problem while on the job there. Off the job, though, he continued to suffer his bouts with migraine headaches and insomnia. Dr. Robert A. Vogel treated O'Donnell and prescribed medication in response to O'Donnell's inability to sleep, as did other doctors. Their prescriptions, doled out to help him sleep and fight painful migraines, made up a concoction he carried around in a duffel bag that rattled, garnering attention from colleagues.

Two years prior, while still in Midland, O'Donnell's prescription drug dependence had grown out of control, and his coworkers could no longer ignore it. On April 30, 1991, at 7:15 a.m., O'Donnell was lying on the floor of Midland Fire Station 8, his bunker gear scattered all around him: huge yellow rubber boots with a reflective strip along the calf; giant yellow pants with thick, red suspenders; black duffel bag stuffed with ratting pill bottles.

"It appeared O'Donnell may have fallen," Captain Larry Capell reported later in a series of written statements.

Capell knelt over O'Donnell, trying to wake him by grabbing one of his shoulders and shaking it. O'Donnell continued to snooze, and Capell kept nudging. "Wake up, Robert. Wake up, man."

There had been little about O'Donnell's performance the previous day to tip Capell off as to any sort of issue or problem. Capell said O'Donnell appeared functionally fine the whole shift, working around the station as if nothing were wrong. O'Donnell was working a shift as a sub for Tim Owens. Capell kept tabs on O'Donnell, though, as rumors had continued to swirl through the ranks. And when Capell arrived at MFD Station 8 around 6 o'clock on the morning their shift began, Captain Ken Carter had told Capell to keep an eye on O'Donnell.

"Captain Carter stated that O'Donnell was rambling in his conversations that morning before I arrived," Capell said. Carter's warning sat with Capell all day. He would later note finding nothing wrong with O'Donnell's speech, driving, or gait the entire shift. "O'Donnell appeared to be fine. He performed his duties without any problem."

Firefighter Bill Wiley went with the ambulance crew on a run, returning to find O'Donnell awake and doing fine. Wiley went back to sleep, finding his bunk in the icy dank tundra that is every bunkhouse in every fire station in Midland.

"He had a good shift and contributed well to the work around the station," firefighter Martin Cory reported, who walked up right as Capell was trying to wake O'Donnell. When they were finally able to wake him, Wiley, Capell, and Cory noted that O'Donnell's condition did not bode well that morning after his shift. O'Donnell's face was flushed, his cheeks and eyes sagging.

"Robert looked like a fighter who had just fought fifteen rounds and was punch drunk," Wiley said.

They continued observing O'Donnell as he made his way to the bathroom in a sleepy daze. O'Donnell tried to comb his hair, failing to even get the comb to his scalp—his fingertips feeling around like masses of rubber and unable to coordinate with each other. "He could not even perform this function," Cory said.

Wiley looked at O'Donnell. "You all right, Robert?" he asked.

O'Donnell could barely reply.

The other firefighters, themselves just stumbling out of their bunks, had yet to put together the several times they had been awakened throughout the night by O'Donnell's tossing and turning. Later, many would report having awakened to the rattling sounds of hundreds of pills in plastic bottles as O'Donnell sorted through the duffel bag he routinely carried. The bunk area of the fire station is a tight space—a single room with Murphy beds lined up beside each other in two columns. Despite the number of personnel who can fit into the bunk area, it is capable of suspending light in a way that makes it feel remote and distant for those needing to catch rest between emergency response calls.

David Pruden awoke in the ice-cold air of the bunkhouse to the sound of O'Donnell digging through his small black duffel bag, peering into it with a flashlight. Pruden did not notice O'Donnell taking any of the pills, nor did he see him the second time when the noise made him open his eyes. Wiley recalled waking up to the rattling three times that night. "He seemed to be up all night in his bed area," Wiley said.

The last time Wiley awoke, O'Donnell was packing his duffel bag of pills and whatever other belongings he had near his bed. He bundled his blankets into the middle of the mattress and lifted the end of the Murphy bed bunk, letting its hinge bend with the whole mess of blankets and mattress disappearing into the wall. There in the pitch-black morning, O'Donnell started to make his way through the wide corridor toward the door.

"He sounded like he was trying to walk through a wall," said Wiley. "I thought he just could not sleep and being dark, he just ran into the wall."

Capell coaxed O'Donnell awake moments later, and helped him to his office, where each captain also has a separate bedroom. Several of the men assisted O'Donnell as he drifted from right to left, falling hard on each leg and causing his knees to intermittently give out.

"Have you taken any medication, Robert?" Capell asked.

"Yeah, but it's, ah, it's—it's prescription," O'Donnell managed to utter through the haze as Capell kept asking questions. To cover his own tail and process a report, Capell called Battalion Chief Eddie Klatt, who drove from Central Fire Station near downtown. With him, Klatt would alert Battalion Chief Gary Chastain and Assistant Chief Ray Sprague. Still talking with O'Donnell, Capell noticed O'Donnell would begin to doze off during pauses.

O'Donnell was finally getting the sleep he had been looking for.

Capell left O'Donnell and briefed Captain Ken Carter on what happened. The two grabbed O'Donnell's black duffel bag and sifted through its depths, cataloging its contents.

Many of the bottles had been filled a week earlier, on August 23, with Dr. Vogel's prescribed medication. Some of the medications included codeine. Capell started counting pills, comparing the amounts missing to what was filled. "From all of the containers we found a large part of the pills missing," Capell told superiors, "a far greater amount than the instructions on the containers said to take."

Capell and Carter started listing the prescriptions, noting the medications as well as the count of pills taken from each container. Left in the bottom of one sixty-count bottle were fourteen tablets. Capell handed the list over to Chastain.

Paramedics Jim McGowan and David Gaylon helped O'Donnell put his shoes on as Chastain and Sprague watched.

"Ray, I want to be checked into Memorial West and have them help me," O'Donnell gurgled as he fell backward on a bed. His words were hardly discernible.

Sprague looked back at him calmly. "I understand, Robert. I know you need help."

Chastain disappeared into the captain's bedroom, lifting the telephone's handle to call Chief James Roberts and Assistant Chief Victor Pena. He returned moments later, and the firefighters were still struggling with O'Donnell.

"We're taking him to Memorial West," Chastain said, referring to a health center specializing in behavioral health treatment for chemical addiction.

CHAPTER 4

The crippling effects of too many pills had been an issue for O'Donnell before. Officials reported at least five instances where O'Donnell had shown up for work under the influence. And just two weeks before being taken to Memorial West for treatment, O'Donnell had stayed at home during a scheduled shift without pay after having shown up for work at MFD Station 3 on April 7, 1991, and having refused to take a drug test.

Battalion Chief Robert White called Chastain at 10:30 a.m.

Chastain had moved O'Donnell to work with White's crew that morning after Captain Russ Conley called Chastain complaining because O'Donnell was five minutes late. Captain Wayne Dorlay needed a driver to replace an injured crewman at Station 3.

O'Donnell found Dorlay in his office shortly after arriving.

"I didn't get enough sleep last night," O'Donnell insisted with the slurring confusion from a stomach full of sleeping pills mauling his tongue. Dorlay did not know it, but O'Donnell's car was still running in the parking lot. O'Donnell had been so under the influence that he had failed to notice the ignition.

Around 10:30 a.m., White's call to Chastain sent a ring through Central Fire Station.

"O'Donnell locked his keys in his car," White began to explain.

"So?" Chastain replied.

"The engine is still running," White answered, noting, too, O'Donnell's trouble with speaking and keeping his balance. "It has been running there for two and a half hours."

By the time Chastain could determine what to do with O'Donnell, he called back to Station 3, where O'Donnell answered the phone, slurring his words.

"Who's there?" Chastain asked.

There was a momentary pause.

"Uh, me. Robert. Uhh. Robert O'Donnell."

"Who besides you?" Chastain countered.

Trey Orozco took the phone from Robert.

"Is there a problem with Robert?" Chastain asked.

"Yeah, I think there is. I think you could say that," Orozco replied.

O'Donnell had asked to be excused from driving the ambulance that day although White had already followed Chastain's orders to take O'Donnell out of service for the day. After hanging up with Orozco, Chastain made his way through Midland's dingy eastside streets, steering the battalion chief's Suburban north along Lamesa Road, with O'Donnell in the passenger seat. The bleary-eyed firefighter fumed angrily, having denied Chastain permission for a drug test. Chastain drove quietly in the awkward silence, heading toward O'Donnell's house.

CHAPTER 5

DOING IT LIVE IN THE PERMIAN BASIN

THE PERMIAN BASIN'S ONLY SOURCE FOR LIVE television news coverage in 1987 came from KMID-TV Channel 2, the affiliate broadcaster of the ABC television network. The station's identity as "Big 2" came with its ability to provide extensive on-location news at a time when this kind of on-the-spot news had yet to fully take shape. The live capability gave the station validity and substantially more vigor than its competitors, making others look obsolete and completely cut off from the action. The alternative meant recording on the scene for B-roll footage of the news items, collecting interviews with sources like witnesses and subject matter experts, and recording standup accounts delivered by a news reporter.

Of course, live capability had been in use already for sixty years. Still, compared to today's technological advancements, which continue to progress exponentially, media consumption and demand evolved at a slower pace. Use of live transmission in local market news, however, came along especially slowly, having been formed in the tradition of a painstaking process of collecting information, verifying it with quotable sources, writing the report, editing the content, and presenting the information that was as factually accurate as the process would allow.

In West Texas, remote live news coverage outside of a studio environment of each affiliate's station had yet to take shape. In 1983, KMID general manager John Foster masterminded a plan to expand their live capability, an extravagant expense for such an isolated television market. But KMID would gain an edge on the market with the addition of mobile production vehicles with microwave antennae that could rise twenty to thirty feet in the air. From that height, the production vehicle could send video back to the news studio.

Transmitting from the field in this way could be a critical advantage for a news station. Although serving small-town West Texas, KMID, as an ABC affiliate, had to compete against CBS and NBC affiliates. KMID's investment would result in leading coverage of the market lasting into the late 1990s. Headquartered a quarter-mile from Midland International Airport and about ten miles west of the city, the station covered the two major urban areas of Midland and Odessa whose collective population equaled that of other major regional Texas markets—namely Lubbock and Amarillo to the north.

The investment would reap rewards in October 1987 on Tanner Drive.

Much of KMID's muscle flexed when Phil Huber, the station's cameraman and one of the first on the rescue scene, gave Glasscock forty feet of microphone wire and a set of headphones. The favor, however, would give KMID a significant edge in coverage. Over the course of the next sixty hours, KMID would drive and overpower other local television news competition with live sounds from the bottom of the well. Glasscock would later credit himself for helping Huber realize the advantage the station would gain in handing over the cable.

Meanwhile, Patrick Crimmins's 10:30 a.m. deadline passed, even though he rushed from the Tanner Drive backyard to phone his story in to his editor Rick Brown. Waiting in the newsroom for copy, however impatiently, Brown had planned to take down the story copy by phone and prep it for layout on page one (known as 1A in the newsroom). Crimmins did what most reporters did at the time—he found the nearest pay phone. The closest public telephone popping into Crimmins's mind was at Dellwood Mall, an enclosed shopping center accessible through a few department stores and two main entrances. The mall had a Dunlaps department store at one end and a smoky bowling alley at the other; Crimmins found one of the bowling alley doors open and waded through the stale smell of old cigarettes and the hall's mechanical grease before he finally located a phone and made his call.

Child Falls into Well
Drilling of Rescue Shaft Initiated

The headline story featured the rescue prominently that afternoon, with a photo of the young mother sitting in the backyard. Her face was forlorn and confused; Cissy was looking back over her right shoulder toward the camera when photographer Curt Wilcott caught her in mid-sentence.

Crimmins was only partially pleased with the story. Dictating the details over a grimy pay phone, he did his best to flip through his reporter notepad, gleaning important aspects to tie the story together. Crimmins's effort—scouring notes for nuggets of detail—typified spot news coverage where time to delivery was more important than content, detail, or sources.

"I got my story for you," Crimmins had puffed when Brown picked up the phone, prompting the assistant city editor's fingers to tap across a panel of keys linked into a video display terminal—a rudimentary artifact from the dawn of the computer age. Not everyone was a fan of this new technology, most notably Editor-in-Chief Jim Servatius. To get the "old-school" Servatius to read his reporters' news copy, someone had to print it out on a dot matrix printer. Like Servatius, some of the newsroom elders hung onto their old typewriters on which they had learned to pound the keys like rapid hammer heads.

Hours later, the first printed news of the effort to save Baby Jessica spun across yards of paper over a grinding press on the north end of the *Reporter-Telegram* building. In another hour, news of the girl's plight and forthcoming rescue would begin slapping and sliding across driveways and front doorsteps as delivery boys and girls tossed afternoon editions from bicycles. Crimmins's breaking news told a short, succinct story:

CHAPTER 5

> An 18-month-old girl apparently escaped critical injury this morning after falling 18 to 20 feet down an abandoned residential water well on Midland's west side.
>
> At noon today, the girl, Jessica McClure, remained trapped at or near the bottom of the well, located in the back yard of a residence at 3309 Tanner Drive.
>
> Rescuers—who included Midland fire and police officers and utility and state highway workers—could hear the girl crying after she fell further down the well during rescue operations about 11:30 a.m.
>
> The child's mother, Reba "Cissy" McClure, told police she was outside watching Jessica and several other neighborhood children play when she heard the phone ring, according to police officers at the scene.
>
> Returning from the call, she saw several of the children looking down the well, officers said.
>
> Police were unsure whether the well was covered at the time of the incident.

Crimmins finished his call with Brown, taking a long breath before hanging up to return to the scene. For all he knew, rescuers could have plucked the girl from the well while he was away dictating his story at the Dellwood Mall.

But back at the rescue site, Jessica had not been rescued.

And no one knew for certain just how precariously her life hung in the balance.

Crimmins hunkered down, preparing for the next day's story with documentation of the rescue. As he worked, he attracted the attention of Corporal Jim White, the MPD's public information officer. The two had a rapport, having worked together, with Crimmins pulling police details from White every day. White revealed he and other rescuers thought Jessica would be safe and under examination outside of the well in just a few hours. Upon hearing the initial reports of Jessica's fall down the well, White had also initially pictured the brick-lined fairy-tale picture of a wishing well. White had only been on the job as spokesperson for five months, but once he arrived and took in the situation, he knew its uniqueness would only intensify and heighten the media's interest in the story. As soon as he arrived on the scene, White recalls, he began managing and coordinating media, and his focus did not make room for much of anything else. "I was hopeful and optimistic that we would get her out, but I didn't get emotionally involved. That's our job," he said.

By 1:30 p.m. on Wednesday, KMID's twenty-two-year-old reporter Rodney Wunsch joined Huber to begin covering the story. Wunsch quickly scribbled notes as updates became available. Going live every few hours and preparing for a 6 p.m. broadcast, Wunsch paced in the Sprague backyard, repeating his news script to himself. Optimism was strong, but first responders began to reveal that perhaps reaching Jessica would be more difficult, laborious, and complex. The shifting dynamics of the rescue revealed themselves incrementally.

Rescuers would stop to address the gathering media, who sat idly nearby without much management or crowd control. The reporters, as they arrived, intermingled with bystanders who had begun to arrive on the scene to explore the commotion in the neighborhood.

"It's a whole lot harder than we thought it was going to be," one of the rescuers explained, updating reporters shortly after Wunsch's arrival. "We can still hear her. We're still giving her oxygen."

"How did the little girl even get in the well?" The question circulated almost immediately.

"We're taking a look at that, but it appears she fell in while playing with some of the other children."

"How big is the well?"

"Eight inches."

"How big is the girl?"

"She's an 18-month-old girl, smaller than the well, obviously."

Obviously.

The reporters could speculate, and while shunned as a practice in journalism, some reporters had learned the tactic of cleverly disguising unverified points as questions posed or pondered by anonymous viewers or bystanders. As Cissy sat in full view of the public at the rescue site in tears and near terror, reporters avoided going fully into a mode of blame and speculation. Instead, details of Jessica's position in the well shifted when it was revealed that rescuers could hear her voice, effectively forcing the focus onto the effort to save her rather than on the causes of the accident. With the commotion and a media style that prioritized the coverage of the breaking news event itself, reporters focused on the drama of the rescue. There would be time for an analysis of what went wrong some other day.

"She's very upset, but they can hear her," Chief White told Crimmins, creating an immediate bond between newsreader and the tiny baby. The thought of a young girl stuck twenty feet down a well with her sweet voice calling from below haunted readers. "Baby Jessica" became a name in the news. The human element in full effect, parents unfurled newspapers, reading intently and beginning to imagine it.

A baby.

In a well.

At the bottom?

How deep does a well reach?

As the rescue efforts stretched into the early afternoon, information about the crisis had yet to fully reach city leaders and elected officials. The city of Midland's chief of police Richard Czech and fire chief James Roberts shuffled papers around a meeting at city hall. The two just happened to be in the same meeting that Wednesday.

Czech, on the job in Midland for only nine months, had begun a reform of the city police administration. Sweeping through the department, he had a plan for wrangling it into accreditation. He would not only establish a set of standard best practices for policing and department policy but also provide the public with the perception that new leaders had assessed standards and made advances compared to those who had come before. Czech's efforts focused on streamlining officers' reports to put them on the streets while also developing firm, concise, and understandable policies and procedures. The fifty-four-year-old's voracious appetite for making the department perform and look good would emerge with time. With his arrival came a steady stream of memos. One such missive prohibited the use of racial slurs by his officers. That became a point of contention among officers, many of whom considered the skyrocketing number of residential burglaries in Midland as a little more important. Czech saw the conduct revision as a necessary move, perceptibly growing satisfied with the results. "I do think there's been a decrease in racial remarks or slurs," Czech told Crimmins in May 1987. Another memo from Czech, coined by the department as "the fat boy letter," rebuked officers whose girth was out of proportion.

Czech's assertiveness accompanied directness—and sometimes he overly narrowed his focus when a goal lay in his crosshairs. In the chief's wake were those who disagreed with his methods. That kind of change is hard for an outsider to exercise in Midland as the pride of the West Texans ran as thick as the Texas crude coursing beneath their feet.

Czech's grizzled features, leaving deep ravines where his puffy cheeks had subsided, accompanied a diligent and deliberate pattern of action. Coming to Midland in December 1986 to take the post held by longtime Police Chief Wayne Gideon upon his resignation, Czech's style of management was much different. Czech's police experience of twenty-three years with a department in Tucson, Arizona—with its 400,000 people—brought constant attention to his desire for changing—or updating—the department.

At the downtown meeting at city hall, Czech and Roberts questioned whether the call was even a real one—they tried to determine whether such a thing would even be possible.

"Our departments are running on a girl in a well," Czech told Roberts. "Is it possible for a baby to be stuck in an eight-inch well?"

Roberts held up his thick, meaty hands, each as big as a bear's paw, forming his fingers into a circle and trying to envision the width of an eight-inch pipe.

CHAPTER 5

"No, there's no way," Roberts affirmed.

The silver-haired fire chief looked at Czech for a moment, pondering the possibility. Trying to consider just what the call could really be about. After all, although Roberts had run on only a limited number of actual fires before he advanced toward administration, he knew many calls to the fire department could turn out to be something else entirely. A call for a man jumping from the roof of the First American Bank building could be a man washing windows on the second story of a downtown building. A call for a multivehicle rollover on the interstate could be a one-car blowout with a driver who did not want to change a tire himself. There was no way to know what a call for a little girl in a well might turn out to be.

But Officer Hall's confirmation from the scene and the continuing calls back to dispatch from Glasscock and Walker confirmed the reported information, and the two chiefs rushed to the backyard, helping to devise a plan.

The backhoe effort someone had contrived might help reach the girl.

"Pull the fence down!" Glasscock heard someone shout behind him.

It became clear, though, in just a few minutes of digging, that the backhoe could be doing more harm than good as the clawing monster shook the ground, worrying crews that the vibrations might cause the girl to slip farther down the well.

Several miles away, Kragg Robinson and Scott Fletcher felt their bodies shake as they maneuvered a 42-foot rig built to gnaw cylindrical holes into the crispy desert crust. Known as a "rathole rig," the large green truck had a diesel engine that roared so loud, Fletcher could not hear Robinson, and Robinson could not hear Fletcher. The two had powered up the heavy auger to dig out pilot holes that would form the footings for 25-foot pillars and support a developing overpass at the intersection of State Highway 191 and what would become Loop 250 on the city's western edge. Robinson's rig was normally in heavy demand in a booming oil economy, used regularly in oil fields for various tasks associated with shallow drilling. Generally, the rigs could be rented for about $250 per hour for projects, but their work had shifted focus as the oil field work ground down to a halt. The developing loop became a sign of Midland's optimism and growth, even in the now stagnant economy. With the city's unemployment rate hovering at 9 percent—while just two to three years prior it remained virtually nonexistent at 2 percent—there were those who still held hope for the future and invested in infrastructure.

When a police officer in a cruiser sped up close to the rig and abruptly parked, they each thought for a moment with a swift recollection of the day's events and lightning quick speculation of just what kind of trouble the other might have stirred up to cause the police to come to their worksite.

"There's a girl trapped in a pipe. We need your rig to dig us a hole next to her," the officer said, spouting off what likely sounded like inane banter from a cop finally gone mad.

"Yep, that's about right," Robinson figured. Sooner or later, he thought, considering all the time he spent digging with the rig, something like this was bound to happen to him.

No one had ever done anything quite like this though—not with there being any remote chance for survival. And none of the men on Robinson's crew would ever be involved in anything like it again. Shutting down the rig and lowering the towering mass to move across town, Robinson thought their effort would likely include rapidly digging out the hole and leaving the scene. Quick and painless.

Getting into the backyard with their large rig, though, would be difficult.

Though several at the scene determined it would be unnecessary to move cable and electric lines to fit Robinson's rig into the backyard, Bentley, the Dimension Cable technician and caver, stayed at the scene. In part to soak in the excitement and also to help if he could, should his skill set be needed, Bentley resigned himself to stand by as Robinson angled the rig through the back alleyway. Robinson maneuvered the rig with its shaft prone along its own spine, parking it within five feet of the well shaft. He could then extend the unit's arm once it was clear of the lines.

CHAPTER 5

By 11:30 a.m., Robinson's rathole rig had plowed through three feet of topsoil with ease. The fine sand granules milled along the rotating shaft of the 36-inch bit, spitting out the matter as productively as a mighty fire ant. Unlike the backhoe, the rathole rig was making quick work of the backyard mission. As the bit spun, it accumulated cubic yards of dirt with its massive, ribbed blades spinning like a corkscrew. Robinson's machine would chew the earth before he would pull the bit from the hole. Using the controls, he would shake the bit clean of dirt and rock and return to the hole for more digging, letting the bit submerge once again. Robinson continued digging and put the full power into the bit just as a hard grinding sound overpowered the mind-numbing drone of the rig's engine. He knew the sound. It was one like many others before, and moments before the police officer stopped him at Highway 191, he had been working the rig through just such a layer of hard rock.

Hoping for soft rock, which might easily break down into a fine dust like busting plaster, Robinson instead encountered a hard conglomeration of naturally cemented rock—caliche. The desert earth of the Permian Basin had once been a seabed collecting the parched remains of a dying ocean. As the sea receded, it left behind a drying mixture of sedimentary rock that became hardened and covered, over centuries, with two thousand feet of blowing dust and dirt. Records of water wells throughout the Tanner Drive neighborhood development, Permian Estates, show a consistent soil profile, with contractors recording various and specific types of soil at various and specific depths. Their striations are somewhat uniform with two to four feet of soft topsoil followed by fifteen to forty feet of thick sections of caliche rock.

Hard. Thick. Like cement.

The screech of metal grinding on the hard granite-like rock sounded like fingernails scraping down a chalkboard—amplified fifty times and pumped directly to the eardrum through powerful headphones courtesy of the KMID-TV crew. Besides being a painful annoyance, the sound was a sign that something was very wrong.

Robinson's stomach sank.

The idea that this rescue would be over in another hour or two vanished.

"What do we do now?" Glasscock muttered.

The grinding bit knocked the wind out of the rescuers. What looked like a quick jaunt underground was suddenly an exploration into the unknown, where a child might soon die.

Robinson could only keep drilling, switching out the bit for a sharper, more powerful one while crews tended to the other, preparing it for another stab at the rock. For hours, no one could do anything but listen and watch—listen to the drumming rig and watch it drill and struggle. The scene slowly appeared more like a staging area for a new construction project.

As Robinson dug away at the secondary rescue shaft, drilling hard in the shaft parallel to Jessica's, Bentley the cable technician overheard the plan to dig a rescue tunnel from the rescue shaft to the primary one. The dig would require experience, and Bentley knew his experience in caving would be key in the rescue.

Fire Chief Roberts bent an ear toward Bentley.

"I explore caves as a hobby, and sometimes we have to dig in those caves to break through to other areas," Bentley explained.

Going down into the rescue shaft would require a man of a different kind of strength and ability. The close quarters meant sweat combined with heat and fine dust that would make even the most experienced driller feel trapped and claustrophobic. If someone panicked at the bottom of the shaft, two people would have to be saved rather than just one, and so far, rescuers did not know with any certainty that saving the first victim was even possible.

"Great!" Roberts replied to Bentley. "You're the first one going down to dig. We appreciate your help."

The scene in the backyard and adjoining alley began to morph. The need to manage the unruly crush of onlookers and would-be volunteers was beginning just as the coolheaded facade of the chiefs was beginning to crack. Much of the angst appeared on the faces of the administrative leadership of both the police and fire departments on the scene. One might have assumed that the burden was due to the task at hand—conquering the thick striations of rock, fighting through them, and getting to the girl—but in fact, it was the growing disorganization of the scene, the management of the volunteers, and the administration of the rescue itself that started to put the chiefs and their respective teams at odds with each other. And others saw it.

An onlooker watched as Czech and Roberts argued over just whose leadership took center stage.

"They were going back and forth," said the onlooker, remembering Czech getting a little hot over Roberts' taking the lead.

For the fire department members, their role was clear in their minds, and it went well beyond extinguishing fires and rescuing kittens out of trees. They were trained in rescue and understood the mechanics of extraction from uncomfortable and dangerously close quarters.

Their skills brought them into the twisted metal of vehicle collisions and their training prepared them for roles in large-scale and complex management of disasters. Small and urban though it may be, the Permian Basin region is home to one of the largest petrochemical hubs in the world. Industrial facilities in the area and the chemicals they used required that local fire teams be prepared to manage unthinkable disasters.

Police, on the other hand, arrived on the scene first. Their skills had similarly prepared them as first responders for any kind of trouble.

This kind of push and pull for leadership authority was not unique to Jessica's rescue scene. Struggle for authority between first responders has always existed in disaster management operations. Here, command and control started to become a major point of concern, as firefighters and police descended onto the scene in rapid succession; their co-ownership of the scene meant that each held command and responsibility simultaneously without a designated voice of leadership. As the minutes ticked by, though, authority needed to be clearly established in a single entity to eliminate confusion and stem the clamoring of ideas about how to get Jessica out of the well.

The need to establish a fully formed and centralized command grew, and the lack of one began to make the problems at the scene grow worse. With a massive hole in the ground forming the rescue shaft, two men at a time working with noisy equipment hammering into a diagonal rescue tunnel, and an array of volunteers at the scene, the likelihood of an accident increased dramatically. With the rescue shaft plunging twenty-five feet underground, its opening would eventually be surrounded by tired rescuers shuffling about. As the assorted participants in the drama grew weary, accidents were bound to happen if the leadership failed to get a grip on a centralized command to manage resources.

Accidents in industrial environments, especially oil field facilities, were a regular occurrence. Studies by the US Department of Labor show vehicle accidents barely register in the top five of the main causes of on-the-job injuries in industrial areas. Instead, chief among the causes of work injuries are machine entanglement and strikes by falling objects. Number two on the list: slipping and falling. A slip on the scene here began to take shape in the minds of those who were more risk aware. One bad move near the rescue shaft would result in embarrassment at best and catastrophe at worst. More concerning, though, was the additional challenge the rescue would become in an already tenuous operation to save Jessica. The threat of a man falling into the rescue shaft began to get very real, especially as rescuers began to stay on the scene too long without accounting for their own fatigue. This type of risk is documented clearly by federal regulators. At the top of their list of causes of major work-related accidents: overexertion.

To manage the scene and limit the overall risk for accidents—as well as the factors that lead to them—incident command posts became a primary practice; but in 1987, their standardized use was not widely prioritized. While it is today considered an initial step in emergency management coordination, there was no such structure in place in the backyard of the home on Tanner Drive. The potential for other tragedies grew as the decentralized management corralled an expanding set of professionals and volunteers who were becoming more emboldened with every passing minute. As more volunteers pulled their vehicles alongside streets before getting out to navigate their way through the alley to the backyard, the foot traffic on the block increased steadily.

Seventeen years earlier, on September 20, 1970, a flicker of flame sputtered to life on Fish Ranch Road in the hills near Oakland, California. The summer months had been scorching hot. The parched grass quickly caught fire and spread as heavy winds pushed the blaze into the steep hillside of the San Francisco Bay. Chronicled in his book *California Aflame!*, author Clinton Phillips says the flames erupted to quickly engulf 36 homes in less than two hours. Another 37 homes were damaged, Phillips notes, and 230 acres of prime watershed were consumed. But the fires continued. For 13 more days a series of fires sprung to life. More than 720 fires then began to sweep through California's southern counties, burning more than half a million acres and wreaking havoc as hundreds of firefighters and emergency responders toiled.

Several of the associated fires were vast and far-reaching; 32 of them burned more than 300 acres each—considered large wildfires. Killing 16 people, the fires cost the state $233 million. Estimates completed more than 30 years later put the cost of the fires at around $1.25 billion.

And no single body existed to establish control and coordination. No one could account for the management of the 19,500 professional firefighters from 500 separate departments. As each group responded on its own, calling for resources and assistance based on its autonomous observations, skills, experience, and needs, the action turned into a frenzy of disorganized chaos. Together, the response formed a sticky honeycomb of confused communication through a variety of disjointed radio frequencies and non-uniform terminology.

Two months went by following the wildfires' foray when California's Secretary of Resources Norman B. Livermore appointed a twenty-one-person task force to review the operations. For two years, the group considered plans for how to mobilize and respond when large-scale disaster bore down on a community. After all, the onslaught of management needs was not limited to the disaster itself as much as it was for the organizational design of those thousands of firefighters and emergency response groups.

To organize six identified areas of dysfunction, the taskforce developed a program using computer technology and a coordination system known as Firefighting Resources of Southern California Organized for Potential Emergencies—FIRESCOPE for short. Having identified a lack of organization, poor on-scene communications, inadequate joint planning, limited resource management, and other issues, FIRESCOPE became recognized for having given birth to the first incident command system. It established a command center with a centralized authority to which first responders could turn for guidance, leadership, and management that would direct shifts of manpower and make adjustments to account for the exhaustion of human resources.

Still, by 1987, only sparse training existed to prepare local municipal emergency management authorities for the evolved protocol set forth by FIRESCOPE with its centralized point-of-command structure. So, the Jessica McClure site began to grow out of control as the lunch hour approached, and it was in serious need of a more robust plan of attack and operational management setup.

While the neighborhood's residents started making initial inquisitive probes of the site from the back alley and adjacent backyards, reporters like Crimmins started gathering more details to continue their coverage. Although the site had yet to be brimming with onlookers and volunteers, local reporters were swarming. For Crimmins, who had made his way back to the site after submitting his first article, there was plenty of time to nail down softer news angles and add more color for his next article. His story would not be due for more than twenty-four hours.

CHAPTER 5

As the rathole rig tore through enormous chunks of rock, struggling to make its way to the twenty-seven-foot target depth, Crimmins and others began interviewing rescuers as soon as they emerged from the backyard. Prioritizing the development of a story that went beyond the hard spot news on this first day fell to the wayside as updates continued spilling over the fence that separated the reporters from the rescuers. What emanated from the rescue scene was a constant churn of details featuring the mechanics of the rescue. Environmental news of the growing rescue operation—details about the scene, the number of rescuers involved and how the process seemed to be progressing—came easily for reporters who could witness the events as they unfolded. Still, their news angles painted a picture of what was supposed to be a quick rescue. It appeared initially that there would be no time to develop softer angles with colored narrative from neighbors and friends who knew the McClure family. There would be no need to develop investigative stories on how the children came to be in the house serving as an unlicensed daycare. By the time those questions could be asked and answered, the rescue would be over and done.

Still, as the hours went by, rescuers remained upbeat and indicated that the rescue would be over shortly. As Crimmins pushed Czech for his view of the progress, he found one indicator that revealed cracks in that short-term projection.

"It's a whole lot harder than we thought it was going to be," Czech told him at 1:20 p.m.

Several hours had passed with first responders still figuring out just how they were going to manage reaching Jessica. Another concern was that no one knew where Jessica's father was.

Chip McClure was hardly yet a man. He certainly had yet to become a seasoned parent. He had been a father for only a year and a half and was still working as a manual laborer. He had managed to get a deal with the owner of a local sporting goods store called The Sportsman's Den, where he had been promised a retail job. Although he had hoped to begin in August, Chip was still waiting to start the job at the store. He had looked forward to transitioning to it from the work he currently had with a painting crew.

"We didn't know we were poor," Chip said of his situation as a young dad and the reaction people had to his life with Cissy and Jessica. They lived in an apartment as a young family but, he said, he grew up with an entrepreneurial mindset and had enjoyed a relatively significant level of financial comfort in his youth. According to Chip, his father benefited from a share of oil royalties and a mid-level management job at a telephone company. He said he appreciated what benefits he had from his dad's finances, which included motorcycles and a significant collection of firearms.

"Money was never an issue," Chip said. "And he [Chip's dad] continued to help us financially after Cissy and I were married. We weren't living paycheck to paycheck."

He and Cissy had dropped out of Midland High School, though Chip admitted a new fondness for high school compared to middle school. "By the time I got in high school, I kind of found my groove, and was a little bit less of a nerd, and I kind of liked high school. But being pregnant, she [Cissy] really was having a rough time. I didn't mind doing the work. I was smart. I was capable."

Although Chip said it was a joint decision, he felt coerced and cajoled into dropping out of school. After getting pregnant, Chip said, Cissy grew jealous of his time at school around other high school girls. Afraid of their allure, Cissy pushed him to leave school, Chip claimed.

Cissy's family, according to Chip, took the news of her pregnancy hard. Devoutly religious and followers of the Church of Christ, her family found it difficult to believe their daughter would become pregnant before marriage. Amid the strain, the young couple kept their relationship going before moving in together and getting married.

Chip said, "I took my GED without even studying and passed, got my GED and . . . just went to work at the same time. It was the right thing to do."

About two years later, he was working as part of a crew painting large apartment complexes when he heard news of the rescue of a little girl in

CHAPTER 5

Midland on the radio around 10:30 a.m. He did not think much of her prospects for survival.

Chip would later say that at the time, he actually had no idea Jessica was even at his sister-in-law's home. When news of the still anonymous West Midland rescue reached him, he shrugged it off.

"Those poor people," he thought while still working on the painting job.

At 12:30 p.m., Chip chomped down on lunch with three others on the house painting crew at an Arby's restaurant on Midland Drive—just two miles from where his daughter was wedged in a well.

A pager tucked on the hip of one of the crew members started beeping for his attention, and while he wandered away to look for a pay phone, the rest ate in a half-hearted grumble over the interrupting call from the crew's boss.

When the crewman came back, he could only bark out quick orders in a hasty huff.

"Let's go!" the crew's boss said to the group.

As the crew argued over half-eaten roast beef sandwiches and salty fries, the crewman huffed again. "Let's go right now. It's important," he ordered, prompting the men to return to the company vehicle in silence.

When the crew returned to the office, the wife of the company's owner met them at the door. Her face was stark white. Chip figured for a moment that he was about to be fired.

"Chip, I don't know how to tell you this," she started. "Your daughter fell in a water well. A police officer is on the way to pick you up."

Dispatched to bring Chip to the scene, Officer Paula Bynum steered for the office of McClure's supervisor. She popped out of her vehicle, a barrel-chested woman with a deep voice belying her short stature.

"Chip, come on," she said. "Let's go."

The confusion of the situation and the abrupt departure sent the eighteen-year-old father into a sort of shock as he tried to understand this reality that had suddenly shifted. Between the news report on the radio, the grim alert from his boss's wife, and the police arriving to take him home, nothing felt real about the entire scenario.

He had been close to death himself as a child. From his own recalled experience or perhaps acquired as his own fully formed memory from several perspectives that formed a kind of family lore, Chip recalls his own near-death experience on the family farm in 1971 when he was just two years old.

Growing up on a farm, kids can enjoy a bit of freedom that comes less from merit and more from the necessity of managing the myriad of chores and upkeep across the vast acreage. Tending to broken fence lines, gathering up eggs from a chicken coop, and feeding livestock builds a sense of independence and freedom. It is a length of rope granted to many farm kids at a very young age. Rod McClure, Chip's older brother by about fifteen years, had a longer length of that rope. The freedom and autonomy of a seventeen-year-old big brother made Rod an alluring figure for Chip, and he yearned for any chance to scurry around the farm with Rod. Feeding the pigs and cows in a pen in the corner of two large pastures, the boys would work their way through the gate near a shining silver stock tank full of water for the livestock. About three feet tall and six to eight feet wide, the tank is a common element on both small and large Texas farms and ranches where natural water sources are sparse. As Rod tended to the feed, Chip climbed a small fence near the tank leading to the pens. No one knows exactly what happened next. Whether Chip slipped and fell or miscalculated a jump from the fence into the pen, it is clear that his young body did not have the height to find his footing when he landed, and he splashed into the water tank.

Rod found Chip floating face down in the tank's murky waters.

When he did find him, he quickly pulled the boy from the water and alerted their mother, who, according to Chip, became uncontrollably hysterical and helpless. The ruckus alerted the rest of the family—Chip's sisters—who came running. While Chip's retelling includes attempts by the family to use CPR, its widespread use and mainstream emergence would not occur for several more years. Still, the family managed to get Chip's lifeless body to the family car. His mother, however, was paralyzed and so overcome with fear she had difficulty reacting.

CHAPTER 5

"As the situation became more and more desperate and their repeated attempts to get her into the family car failed, it seemed this wasn't going to end well," Chip said. A family friend who lived nearby, Geneva Giles, suddenly and inexplicably arrived for an unannounced visit to find the family in a panic. As she wheeled her car onto the property, she hurried to the group huddled around Chip turning blue and still lifeless. As Chip tells it, Geneva took control, taking up Chip's body, placing him in her vehicle and speeding off for a hospital. She somehow maneuvered quickly enough that the hospital staff was able to express the water from his lungs and bring him, coughing and gasping, back to consciousness.

"Obviously, I was too young to remember, so I'm retelling this story as I grew up with it," Chip said, recognizing Giles as a sort of guardian angel who still watches over him. Chip, despite his youth, still felt as though he was living on stolen time from that brush with death.

Experiences like these somehow concretize themselves despite the nebulous haze of early childhood and yet somehow, from a collective, form a truth that makes us who we are. Those experiences in some way can stick with people and inform their response at a particular moment. So, by the time officers could ferry Chip to Tanner Drive, the reality of the scene with police cruisers and media caused a momentary detachment from the reality of what he was facing. The flurry of people on Tanner Drive along with the first responders fed a growing and unsettling emotion. He attempted to take in just what was going on around him as rescuers flurried and police lay on their bellies beside the well at his sister-in-law's house.

According to Chip's memory of the moment, he suddenly felt a hand on his shoulder and heard the sound of a deep voice.

"Don't worry, son. We'll have her out of there before too long," said Chief Czech at his side.

Despite his best effort, Chip fell into a daze, questioning the operation and convinced his daughter was not actually trapped in the well and was instead somewhere playing in the neighborhood.

He could not wrap his head around the situation fully until Glasscock placed the headphones on his ears to let Chip hear his daughter's voice coming from twenty feet down the well. Getting a view of Jessica in the well also confirmed it for Chip, though not much could be seen from the camera, which was traditionally used by the city of Midland to inspect the seams on sewer pipelines.

Little could be made of the images the camera captured. A swath of bamboo leaves obscured most of what they could make out from the top of her scalp and head. Jessica's playmates apparently dropped the leaves into the well after Jessica slipped down the shaft. As it was explained, the children had been dropping objects down the well as part of their playtime activities in the yard when Jessica fell into it.

Jamie Moore, who lived in the house with her husband, James, returned to the scene after classes. She thought the wellhead had been covered, she said. She assured her sister and brother-in-law it had been that way, though accounts vary on whether the well opening had been covered with a rock or a flowerpot.

"I'm so sorry," Jamie Moore told Chip and Cissy. "I've always kept it covered. I don't understand."

Midland police detective Manuel "Manny" Beltran joined the effort at the well side, peering into the well intermittently with a pair of binoculars and listening on a set of headphones to the whimpering sounds from Jessica. Reports of the voice struck rescuers, their emotion filtering to reporters at the scene.

"Hey, what about hypothermia?" Glasscock heard as paramedics continued assessing needs.

"Well, how cold is it down there?" Glasscock asked.

The question was a good one, and with a little thought someone was able to round up a thermometer in the neighborhood. The lightweight plastic gauge took the shape of a small owl, its round eyes staring back at Glasscock as he helped rig it to send it down the hole. Taping a metallic Zippo lighter to it for weight, and tying it to a length of line, Glasscock

71

slipped the little temperature gauge down the well, placing it just above Jessica's head.

Beltran and Glasscock stared at the gauge over the next two days, keeping track of the temperature, which hovered steadily at 65 degrees—uncomfortably cool for extended exposure, but not life-threatening.

"Doesn't Southwestern Bell have a machine that pumps warm air down into manholes in the street?" someone pondered aloud.

Before Glasscock noticed anyone make the request over the radio dispatch, the giant air-blowing machine's bouncing orange lines found their way into the backyard, crawling over the Spragues' fence. Still, Glasscock and many of those working at the scene did not realize the reach of their power as all eyes in Midland began to turn their way.

"Everything we asked for, we got in minutes," said Glasscock.

Rising through the ranks of the MPD and taking on a variety of training programs, Glasscock had not been prepared for a rescue under these circumstances. Without a single strategy, Glasscock and Beltran decided to take whatever leadership they felt necessary. Even as Southwestern Bell's heating ducts stretched to the scene, skirting the bamboo shoots and a large agave plant along Maxine's fence, rescuers had to determine just how the heat would reach Jessica. The warm air would certainly dissipate and rise back toward the men bent over the well. A slew of wires and tools were strewn around the hole by the afternoon, and a group had started examining how they would implement the heating system in the well shaft. Fashioning a twenty-foot section of PVC pipe into a vent, Glasscock hustled down the alley probing the fence line and garbage bins for a connector from the heating hoses to the PVC pipe. Traffic was already a problem as police cars, utility vehicles, and ambulances piled into the alleyway. Officers standing guard at each end of the alley blocked the way for more vehicles using frantic hand signals and traffic cones.

Glancing past the cones at first, the idea of funneling air clicked into his head. He grabbed one of the cones, pulled out a pocketknife, and sliced the black rubber base from the cone before rushing back to the well.

Duct taping a heating hose to the cone's base, Glasscock tucked the small end of the cone into the PVC pipe. A crossbar held the system in place, balancing it on the well's brim.

Meanwhile, though much of the work was assessment, firefighter Felice's internal motor revved with nothing to do but look on while Robinson struggled farther. Felice felt anxiety build as the drill pushed its way through the earth. He looked around, assessing the rescue and imminent needs from his own perspective. As Robinson continued boring the rescue shaft—pulling the drill to the surface occasionally, spinning dirt from the bit—Felice watched and waited.

"It was sort of a hurry up and wait situation," Felice said.

He had watched as the backhoe fought with the caliche rock earlier. Such setbacks did not rest well with Felice, whose unsettled energy sent his mind thinking ahead. If drillers were going down, hammering into the rock, drillers would need air. Although lugging a twenty-pound air tank would be easy—Felice easily managed the weight of an air tank along with another twenty pounds of bunker gear at fire scenes—there would be little space for a man and a drill. Having an air tank would cramp space even more. Assistant Fire Chief Dean Williams and Felice talked, looking on and figuring out a way to keep from lugging a heavy and bulky air tank down the shaft.

"We really need some sort of breathing air system," Felice told Williams. "I'd really like to have the rescue truck out here or some sort of breathing system where we could hook into cascading pressure lines."

Williams did not pause. "Okay, I'll get it here for you."

The intuition of those implementing a makeshift life support system around the girl seemed to be some of the only progress made efficiently while Robinson continued drilling. His rig still fought on through the rock at 3:30 p.m. as Crimmins caught up with Czech for another update.

"How's the progress on the drilling, Chief?" Crimmins queried as Czech sauntered off toward the shade of a tree away from the scene.

"Slow," Czech curtly replied, not stopping to chat.

CHAPTER 6

TOO FAR GONE

FOR FIFTY-EIGHT HOURS, ANDY GLASSCOCK—HARDLY A hardened police force veteran—worked on adrenaline. It coursed through his veins, fueling an attachment to the wellhead, where Jessica's voice fell delicately on his ears. He could just make out the soft and sweet chords of a baby's developing vocabulary.

At hour fifty, his body caked in the fine dust that billowed from the rescue shaft and shaken from the bodies of drillers who emerged, Glasscock stayed near the well, wearing headphones, listening to Jessica. He broke his exhaustion into intervals, leaving the well for an occasional interview with reporters clamoring for an assessment.

For too long, perhaps, he had remained the focus of the camera's lens, so much so that once the rescue's end came, he clamored for attention the same way media had for his quotes in the hours before. He would spend years reaching for the spotlight, holding onto whatever shreds he could harvest off the McClure family as their own media fanfare grew.

On April 7, 1988, a month before Midlander George W. Bush and a handful of investors would take over the Texas Rangers baseball franchise, Glasscock stepped into Arlington Stadium with Jessica McClure and her family. The police sergeant had become a trusted liaison for the family, working as a go-between for the media and the family, making sure the family's exposure was not developing into a safety risk. More than anything, though, Glasscock served as a buffer between the family and the anxious world waiting for news on Jessica.

"Jessica, even as a baby, recognized him and, you know, was very comfortable with him. They all had formed a special bond," said Glasscock's wife, Lynne. She watched the developing attachment, describing it as a security blanket for the McClures. Looking back, she said, maybe it was a security blanket for Andy. The police department sanctioned the security events, allowing him time off from work.

"There wasn't a lot of grief given to him," Lynne surmised, admitting the operations were perfect public-relations tools for the department. Putting Glasscock out front as the man of the MPD meant that a recognizable face greeted crowds longing for the affectionate grin of Jessica McClure. Meanwhile, fellow officers were left behind to play by the book and do their jobs like every other cop, while Glasscock had found a way to skirt some of the rules. He admitted himself feeling a self-inflating glow of importance, having taken a center-stage role in the continuing Jessica saga.

Glasscock was on the road with the McClures when they traveled to Lancaster, Ohio, for a benefit; to Fort Worth for a horse show; to Arlington, Texas, for Jessica to throw out the first pitch at a Texas Rangers baseball game. "Look at me. I'm somebody," he explained of his thoughts at the time as he would leave for a McClure-related event when fellow officers stayed behind for their regular police shifts.

By that April, Glasscock's wife felt pushed away—just shy of two years into their marriage. As the press coverage swelled and Glasscock was interrupted at dinners at local restaurants to shake hands with gleeful passersby recognizing the high-profile police officer, Lynne felt the constant attention as a burden—though her husband seemed to relish it.

The attention and praise seemed to reach its peak that day at the Texas Rangers game at Arlington Stadium as Jessica took the field for the game's ceremonial first pitch. The media swarmed to get photos of Jessica, wondering just what the little girl would look like when she emerged onto the field. As the cameras began clicking, their eyes narrowing focus, Glasscock took note of the camera angles, watching the flashes from the crowd. In a single moment as Jessica prepared to make her pitch, rearing back her throwing arm, Lynne felt her husband's hand push against her shoulder, moving her away and himself into the media viewfinders.

"We're gonna have some trouble here," Lynne thought, feeling her husband slipping away from her. "That kind of behavior continued. He would step over here into the spotlight, and I was on the sideline," she said.

On September 18, 1988, Lynne's first child, David, lay in a crib nearby at 6:45 a.m. when the phone rang. He had been born ten days earlier and was Andy Glasscock's third child. He had two from a previous marriage. The day before, Hurricane Gilbert had swept over Mexico's Yucatan Peninsula, ravaging the developing resort areas of Cancun and Cozumel.

Lynne listened to Glasscock as he coaxed himself awake, on the phone with Midland County Sheriff Gary Painter. Painter was asking for help getting rescue supplies to stranded victims of the hurricane's wrath in Mexico. He and US Marshal Phil Maxwell prepared a 737 jet for takeoff from Midland International Airport. "We've got a plane sitting on the runway. Can you go?"

"Sure, I'll be there," Glasscock affirmed.

Lynne, trying to wake herself, rubbing her eyes, asked with exasperation, "What are you doing?"

David whimpered quietly in his crib. Lynne, exhausted from late-night feedings and keeping up with making sure she did not do anything stupid—the emotions and nervousness of a first-time mother—was dumbfounded by surprise at her husband's news.

"I'm going to Mexico to help with the relief effort," Glasscock said deliberately and without a ping of uncertainty.

The 737 would deliver 10,000 bottles of a green electrolyte drink to stranded hurricane victims. They also packed a barbeque pit. Glasscock calls it his "lesson on how to go to Cancun on twenty dollars."

"That's all I had in my pocket," he said with a guarded laugh in a May 2005 conversation. "My wife put up with more crap from me than any human being should have to," Glasscock said, admitting years later to his unyielding grasp on the spotlight. Eventually others began to agree with Lynne's assessment of who Andy had become.

"You're an asshole, Andy."

The words, although he had heard them before, hit the pudgy cop hard. This time, they came from the marriage counselor Andy and Lynne had begun seeing on the recommendation of a family friend. Glasscock stopped talking, taken aback for a moment. The glazed-over smile Glasscock had mastered for TV interviews slowly fell from his face. The rosy apples of his puffy cheeks drooped, and his chipmunk teeth disappeared to an expressionless surprise.

"You have shut Lynne out for so long, she has put up a wall against you," said the couple's counselor. "Your marriage may not be able to be saved."

The counselor's point struck a chord. Few others had gotten through to him, but as Lynne and Andy both agreed years later, this third party's voice cut through the void between the two. Glasscock finally saw, he admitted for the first time, what others close to him could see. From afar, he was a force for good, heroic. In everyday life, however, he had become self-centered, unable to go about a normal life as a father and a professional lawman.

As the days following the rescue saw the media spotlight beginning to dim, Glasscock had fueled his need for adrenaline and attention with any form of excitement. He took every interview. He sped his way into car chases. Donning a wetsuit, he plopped into Wadley-Barron "lake" for a submerged vehicle after it reportedly ran off the road and sank. And he took the necessary steps to become a bomb technician, enrolling in an explosive ordnance disposal training at the US Army's Redstone Arsenal in Alabama. Andy's progression into riskier law enforcement arenas escalated.

"In retrospect, I think it was another way to find his adrenaline rush," Lynne explained years later in her assessment of Andy's approach to his career as the spotlight around Jessica's rescue dimmed.

With the explosives training, Glasscock not only would perform in various bomb disposal operations but also would appear in court testifying in various cases where he had been the bomb tech on the scene. Following a 1997 standoff with Republic of Texas separatists in Fort Davis, Texas, which garnered international attention, Glasscock conducted a review following the arrests of the group's leaders. According to the criminal charges against the group, law enforcement identified a string of pipe bombs in creek beds and trails leading to the group's grounds, which they claimed to have been the headquarters of the sovereign nation of the Republic of Texas. As dictated by protocol, Andy blew up pipe bombs, set off booby traps, and cleared the creek bed of other explosives. In a video of the effort, Glasscock can be seen hopping through the creek bed from rock to rock like a child, wearing his thick, protective bomb tech suit. The video, presented to a jury in US District Judge Royal Furgeson's court in Midland during the summer of 2001, showed Glasscock setting off pipe bombs potentially stuffed with nails and shrapnel meant to maim. The Republic of Texas members responsible for the pipe bombs claimed they were intended to scare off potential trespassers and did not contain explosive material at all. In any case, the bombs were destroyed by Glasscock in the name of public safety.

Rarely did Lynne know exactly what her husband had been getting himself into, but it was not all that difficult to learn of his exploits and hear news of his police work. She just had to click on the TV news, she said.

"They would call him, and there he would go. He was getting involved in all kinds of stuff," Lynne said. "As media arrived on the scene, so too would Glasscock front and center."

Glasscock's breakthrough realization in the early 1990s, precipitated by the counselor, prompted Glasscock to return to his office the next day and remove every scrap of Jessica McClure rescue memorabilia. He stood there, looking at the world he had been living in, walls covered in rescue clippings and plaques.

"I started looking at all the stink I had done to Lynne," he said.

CHAPTER 6

Although in that moment he started to let Jessica's rescue slip from his daily thoughts, he could not let go completely. He continued looking for fun and excitement to fill that void left behind by the rescue saga.

As Glasscock worked to let go of the Jessica adventure, he knew another key player in her survival was not letting go. Firefighters and police officers usually know each other well, complementing each other's tasks on emergency scenes and depending on each other's roles. Robert O'Donnell, unlike Glasscock and many other rescuers, had yet to let it all go. Glasscock started to avoid him whenever he saw him.

"All Robert wanted to do was talk about Jessica," Glasscock said.

CHAPTER 7

OIL TOWN GOOD OL' BOYS COME TO THE RESCUE

Prospects for Quick Rescue Shine Bright

SOUTHWESTERN BELL TECHNICIANS PUMPED WARM AIR down the well shaft to where Jessica was suspended in an unknown precarious balance. Glasscock and Beltran watched the temperature gauge start to rise. As 5:30 p.m. approached and Robinson's rig worked through the punishing rock, Roberts and Czech continued painting an optimistic picture for media. Reporters began gathering in Maxine Sprague's yard, each putting a chin along the fence to get a view of the rescue's action. Other onlookers and bystanders assembled alongside media just to get themselves a view as concern in the neighborhood swelled. People stopped and stared as reporters prepped their broadcasts, pacing back and forth in the yard and trying to memorize their words for the camera.

The waning light highlighted an uncomfortable fact: Jessica would be trapped for a much longer period than initially imagined. Still, it seemed that after the parallel shaft were dug to completion, trudging diagonally five feet would afford quick access to the little girl. Despite the slow pace of the auger bit from the green machine striking hard against the caliche rock, optimism reigned.

As the daylight faded, one of the Southwestern Bell technicians found Roberts.

CHAPTER 7

"You want us to rig you up some lights?" the tech asked.

Roberts's forehead crinkled. "What are we gonna need flood lights for?" Roberts asked. "We're gonna get her out in just a few minutes."

The layers of rock Kragg Robinson fought through with the rig—which was powered up to the point of needing a change of drill bits every few minutes—foreshadowed the looming trouble that might be encountered when it would be time to start chipping away at the horizontal rescue tunnel. What rescuers thought would be veined striations of rock was actually sheets of bedrock running several feet thick.

Robinson had been expecting it.

"We had already been drilling into that same type of formation, so we had seen everything," said Robinson. "And to tell you the truth, we couldn't do it any faster even though we were going straight down."

The normal rate for running the rig was $250 an hour; Oscar Robinson, the rig's owner, was looking at thousands of dollars in lost drilled time. But he didn't mind, saying simply, "I have grandchildren of my own."

No cheering came with the completion of the parallel shaft. As Kragg Robinson shook off the last bit of soil from the bit, yanking the rig out of twenty-seven feet of empty shaft, crews grew concerned that too much time had passed. In the monitoring of Jessica's voice, anxiety rose in the stretches of her silence. Did the silences portend the worst? Despite calls into the well, Jessica would not respond. Even though they had finished the rescue shaft and started aiming their diagonal shaft, there was no cause to celebrate.

"Jessica! Jessica!" Beltran and Glasscock would yell into the well when she went quiet. Worried the girl was not actually asleep but instead unconscious—or worse—they took turns verifying her condition by trying to get her to call back to them.

Earlier in the day, Cissy emerged from the home and sang into the well, and up from the depths came another singing voice. So, Glasscock gave the tune a shot as well. Clearing his throat, he sent a bellow of notes echoing into the shaft to catch the girl's attention.

"Winnie-the-Pooh. Winnie-the-Pooh. Tubby little cubby all stuffed with fluff, he's Winnie-the-Pooh. Winnie-the-Pooh, Willy nilly silly old bear."

OIL TOWN GOOD OL' BOYS COME TO THE RESCUE

Her voice cried back, gurgling the words through her sobs.

For the rescuers, relief came at the girl's expression of frustration. It was a sign that she likely had suffered few injuries. If she had the energy to respond in frustration, she had the energy to survive. For first responders providing aid, hearing a victim or patient responding to pain or discomfort is part of an assessment. They had no way of knowing the extent of Jessica's injuries. No way of knowing what limbs might be broken and limiting blood flow. No way of knowing whether there were internal injuries causing bleeding. And Jessica's response was strong enough that it settled their knotted stomachs for a moment, allowing them to rest easier knowing there was still enough time to rescue her.

A severe injury would have been cause for critical concern as she would have been losing blood from an open wound or, perhaps worse, have her insides battered with any number of internal injuries. Depending on the extent of the injury, Baby Jessica could have been dead by the time the first driller even pressed jackhammers into action Wednesday night. So, for the paramedics, the cantankerousness was an indication that no injuries threatened her survival.

"If she was bleeding, she would have already bled out," Walker grimly told a *Reporter-Telegram* writer.

Some of the good ol' Midland boys were getting worn down, but Jessica's condition gave everyone a second wind. It kept drillers chipping away, even as the muscles in their hands seized up, their bodies shaken and beaten from the incessant pounding. One round after another at the bottom of the shaft meant another beating, but men were falling over each other to make a descent to the bottom of the rescue shaft to slither along the tunnel and take their turn chipping away little bits of rock.

Others stood by, some idly, while others were given duties like sharpening drill bits. A welder, who arrived on the scene to assist, told reporters with United Press International that teams had sharpened several hundred drill bits, many of which had been dropped off by anonymous oil field hands at the scene.

CHAPTER 7

Wayne Johnson helped with the drill bits. Johnson, the fifty-seven-year-old owner of Johnson Specialty Supply Co. in Odessa, welded tungsten carbide onto bits, enforcing the bits' integrity and lending them several more hours of use before drillers would send them back to be sharpened once again. Johnson leaned over the tailgate of a work truck, melding the carbide material with the bits over a thick iron anvil. The blue flame spit a jet flame onto a collection of bits, one by one.

Johnson pulled a red-hot bit off the anvil, placing it in the prongs of a set of long forceps. As Johnson took his attention away from the anvil, he bent over and dipped the red-hot bit into a vat of oil to temper the agent. As he did so, Chester Massengill walked over to help and placed his hand on the hot anvil where Johnson had just been welding. A sizzling fire grew beneath Massengill's hand, burning him severely. Yelling with gruff, cursing anger, Massengill looked at the seared flesh of his hand before looking back down to see what caused the furious pain. Massengill received second-degree burns to his hand. The slew of volunteers at the scene had neared a state of unruliness, fueling concerns about a more serious accident than a burn taking place—one that might mean the difference in success or failure.

Bentley's carabineers hooked into Robinson's rig well, requiring both men to help drillers descend one by one into the shaft. Robinson's crew, along with a number of firefighter-paramedics, would take turns drilling after Bentley, and switch out in shifts. If it took eight hours to dig almost thirty feet, rescuers estimated it should take only a few hours to blow though five feet across to Jessica—three hours at the very most. The simple math of it indicated that the diagonal shaft should take no time at all.

Bentley hooked in on the rescue line, easing back in the cradle of a rappelling harness. A twenty-five-foot strip of flagging tape marked the length of Jessica's well shaft, pointing the way for Bentley's jackhammer. The twenty-eight-year-old knew how to handle cramped spaces, and he had little concern for how narrow the working space would be.

OIL TOWN GOOD OL' BOYS COME TO THE RESCUE

Robinson's rig cranked up with its roaring, teeth-grinding furor as the gears lowered Bentley. With each descent and ascent, the growing thunder of the Green Dragon's monster diesel engine roared. Each time it sparked, so too would reporters run to the Sprague fence to see which rescuer was next up to go down.

In descent, Bentley watched the layers of hardened sediment give way to thinner sections of soil, like examining a soil sample drilled out of the earth, as he was lowered into the shaft. Layers of the thick rock ran in thirty-inch sheets in some places, while the cemented sediment was fused in striated veins like marbled fat through a ribeye steak. He dangled his way toward the bottom. He reached a gloved hand out to the wall, touching it and brushing at the surface. Still in a Dimension Cable uniform shirt, he angled his bit to the wall, taking aim and laying the powerful chisel into the rock for fifteen minutes. His energy did not wane. Still in shape and physically able to hammer at the wall for a longer session, Bentley knew he had more energy to keep working away at the wall when he felt a tug on the line from the crew above to let him know it was time to switch drillers.

"Man, this is going to take forever with this rock," Bentley shouted as they pulled him out of the rescue shaft, his hearing dulled from the noise of the bit and the Green Dragon's engine. Bentley's hearing would not be fully restored for almost another week.

Little thought was given to liability and legal risk. Few focused on who could and could not volunteer. At the time, a full assessment of the risks involved was of little priority. Without a command center between police and firefighters, confusion mingled with effort. Some questioned just who was in charge—the police or the fire department. Rather than gaining an official method of organizing, rescuers just kept moving and responding as each need arose and made itself known. Perhaps the urgency of the situation had in fact called for each team to put blinders on and forge ahead. However, cracks were emerging in the ad hoc chain of command. Establishing a central figure to manage the scene would serve a purpose, as confusion about the chain of command resulted in mixed messages being sent to volunteers and first responders. Still, there was a feeling among those present that there was no time to organize. "Not the time for meetings," the rescuers said to one another.

Without a central command structure and no real precedent advocating the need for one, the risk of a catastrophe only increased. Firefighters and police said it became clear that, although not a wildfire that could spread quickly, the scene needed to be managed more efficiently as volunteers and first responders weighed in. Enough stories from oil field roughnecks swirled around coffee shops in the Permian Basin for everyone at the scene to envision how easy it would be for a piece of equipment to roll off a truck or for a heavy load to slip from grasp, only to tumble into the wide mouth of the rescue shaft. Or, worse, into the small well shaft where Jessica hung precariously. What then? Someone would surely have to answer questions that would result from such an accident. Years later, such answers would be forced into the public sphere of social media, bloggers, and a twenty-four-hour news cycle. In 1987, though, such an answer would follow only the questioning of a tearful young mother wondering just how those rescuers were so poorly managed that they would be allowed to roam about the rescue scene as they pleased. Those eyes, the sobbing West Texas twang of a young mother, hardly an adult herself, would most certainly not come with a measure of context or understanding. There would be no solace. No one would be accountable.

The fire department felt its responsibility heightened since the scene had become more of a rescue operation; based on their skill and training, they believed they were best positioned to manage the operation. Some of the police felt their role was foremost due to their arrival on the scene early and with an established jurisdiction. Certainly, police and other law enforcement possessed a level of skill and training in various search and rescue techniques, but the MFD also prepared its personnel with a multifaceted skill set that enabled them to conduct the rescue as skilled firefighters. Moreover, the MFD, as a city municipality, cross-trained their teams as paramedics. While the city of Midland is not alone in this practice, not all municipalities work the same way. Fire departments require emergency medical technician basic training. Advanced training as paramedics, beyond the basic EMT certification, requires more than a thousand hours of additional training.

Robinson and his crew of firefighter-paramedics, along with Bentley, were among the first to shoot their hands into the air, volunteering to help relieve drillers at the bottom of the shaft. But could they? Bentley worked for a cable company and would obviously not be covered by city insurance if he were injured. If he succumbed to the elements or became involved in an accident, who would be targeted as the responsible party?

Bentley did not return to drill in the rescue shaft.

Felice followed Bentley into the rescue shaft, hooking into the rigging and taking Bentley's jackhammer. A marathon athlete in his early thirties, Felice manhandled the drill bit. With a physical capacity to handily outpace rival firefighters at the annual Texas Firefighter Games, Felice knew his stint in the well could easily go beyond the fifteen minutes allotted to Bentley's dig.

Felice brushed his hand over the spot where Bentley had pounded for fifteen straight minutes. There seemed barely an indentation. Obviously, Felice surmised, there was work to do, and the system put in place did little to help matters, as the rotation between drillers coming up the shaft, switching out harnesses, and going back down the shaft was inefficient.

"They were lowering buckets down to me," Felice explained. Each bucket—only partially filled with marble-size chunks of rock—was sent back up.

A radio lowered to the bottom of the shaft to keep tabs on drillers' progress only took time away from drilling. Intended as a tool for the crew above to receive active reports on progress, the constant distraction only slowed the drillers' progress toward their intended target. Hoisting buckets up, then lowering them down, Felice had to make sure the buckets did not hit him or the drill. Powering off the jackhammer then powering it back up repeatedly, the precious lost time had to be made up by another driller.

"How's it looking down there?" a voice would call from above over the radio line. The curious officials above would keep asking questions like a concerned mother whose first child leaves home for the first time. "How are you doing?"

Felice's frustration flared, and his only method of blowing off the raging steam was to put full pressure back into the rock. "No, I'm cool. Tool's good," he said, kicking it back on and running with all the power he could.

Felice started to realize the number of chiefs in the mix meant the voices of the volunteers could not be heard. Too much was swirling on the scene. Felice earlier had been sitting in a classroom, learning about incident command. He shook his head and laughed, staring straight at his target and powering into the rock. Running the tool wide open, as he described it, and pushing it as far as it would go, he was unsure of exactly what effects drilling in the hole might have for drillers.

Rescue managers had decided that an allocation of only twenty minutes in the hole at a time would efficiently allow close examination of their conditions. After twenty minutes of Felice running the hammer at full force, workers tugged on his line.

"I'm fine. I'm fine. I'm still going," he replied to the tug in frustration, starting the drill up again. "I'm just getting warmed up."

Then, Felice felt the air from the compressor in his tool give way. Above, someone kinked the hose—the only way to keep rescuers from drilling full force without breaks. At least one person on the scene said it was Roberts who had mandated Felice's transition out of the shaft—keeping a close eye on his men and limiting the amount of time they were allowed in the hole.

Felice would go down several more times to chisel in the diagonal shaft, eventually getting each of his shifts extended. In all, he estimated he spent three hours in the rescue tunnel—much of it drilling and other times recovering while another volunteer hammered at the rock.

CHAPTER 8

FIFTY-EIGHT HOURS OF DAYLIGHT

Night and Day Collide at the Rescue Site

NIGHT DESCENDED ON TANNER DRIVE, AND LARGE floodlights arrived at the scene, suspended atop transportable racks powered by rolling electric generators, while crews continued drilling their way diagonally from the rescue shaft toward the well shaft. Their best estimates told them their trajectory would put them just inches below the girl, but hushed worries plagued leaders as they pondered the possibility of plunging directly into Jessica's exact position. The plot of their course could be completely thrown off if she managed to slip any farther down the well shaft, so further worry about their position gave way to the notion that she could die from exposure or unknown injuries. Or that she could slip down the well to a position from which no one could reach her. The rush was on, and it was clear the volunteers would not be able to make their way through a crowded residential backyard and alleyway without shedding some light on the operation.

What had become the onset of darkness suddenly exploded into daylight as crews kicked on the set of towering floodlights. The lighting had a major effect on those moving around the scene. The night's absence combined with the flurry of activity around Tanner Drive eliminated the innate sense of the passage of time. Glasscock and some of the other rescuers said it altered their senses and made checking the time essentially meaningless. The time of night and day became a set of nebulous numbers with no context. Day, night, and the hours on the clock became irrelevant, they said. Instead, time became a calculation figured by hours and minutes counting forward—starting at 9:30 a.m. that morning, the moment Jessica disappeared down the well—and clicking on hour by hour.

At hour twelve, approximately 9:30 p.m. on Wednesday night, work carried on like it did during the day. Robinson's Green Dragon belched to life with each passing driller. The drillers rose from the depths choking on dust, the finest chalk particles rising into the air. The faces of the men were clear and clean where their oxygen masks protected their airways and eyes. Their hair, clothes, and all surfaces not covered carried a thin layer of fine white powder. It wafted in a plume into the night.

Bentley, though he would not go back into the well, stayed at the scene to assist—untethering one driller before tightening the next one into the harnesses properly. He clipped the drillers in and out of their rigging, asking each as they came out how the progress looked, though many of the men struggled to answer through hacking and coughing as the fine dust settled in their lungs.

"It's going to be a long night," Czech said around 9 p.m. after wandering near the reporters for a break.

Several of the rotating drillers already looked exhausted and ghostlike, covered in the chalky white caliche dust. They were gaunt and stared distantly looking for water or a soft place to lie down. Despite their strength, these strapping oil field drillers, accustomed to dirty work and heavy equipment, were overtaken by the condition of the rock. Crews could see the toll it was taking on the strongest of the workers. The drilling became so draining that some of the paramedics were pulled away to conserve their energy for the moment when Jessica would be able to slip through the diagonal tunnel. It had been decided that a paramedic should be assigned to pull her to safety and to immediately assess the child for injuries. A paramedic's assessment and treatment meant rescuers were viewing Jessica as a patient, caring for her like a victim at the scene of a car accident. Their training would fit the situation like any other rescue scenario. Additionally, they could conduct a rapid medical assessment the moment any part of her body could be reached.

As the drillers continued rotating into the well, many were exhibiting total exhaustion. Some of them were still in the work clothes they were wearing when they showed up to volunteer; some had been off-duty firefighters or utility crews who arrived at the scene to make sure the area was safe to drill. Some came to the surface appearing to hyperventilate, while others only wanted water to quench the parched linings of their throats and splash the dust from their eyes.

"We're going to need more drillers," someone decided.

Finding volunteer drillers would not be too difficult. West Texas is awash in oil field roughnecks just waiting to work, if not already in line at the employment office. Some of them had been turned away from the scene earlier that evening, when rescuers thought Jessica was just a few minutes from salvation.

Charles Boler, who along with his father and several brothers ran Boler Pump Company, had volunteered at the scene when quitting time rolled around Wednesday evening. His offer to help was turned down. By Thursday morning, however, the tables had turned. The troublesome rock had revealed itself to be a formidable foe, and crews started reaching back out for additional help from volunteers.

On Thursday morning, before radio station newscasters prepared for their morning programs, rescuers sent word for more help. On-air radio hosts were already sucking down coffee and voraciously checking through headlines coming off the newswires as the rescue neared its twenty-four-hour mark.

"We need anyone at the scene with any type of drilling experience," said morning news talk radio hosts, kicking off their broadcasts. The message set fire to the streets, clearing employees from some business operations for the whole day.

Charles Boler and his eighteen-year-old son, Ribble, returned to the scene to see whether they could help. In tow were Larry and Floyd Boler, Charles's brothers, as well as Dave Perry, another employee of Boler Pump. They pulled onto Tanner Drive toting their jackhammers early Thursday morning. This time they were not turned away.

The call had gone out all over the state.

Steve Forbes, a firefighter-paramedic in the MFD, also heard the call spill over the airwaves. At 5:30 a.m., he had just finished up a part-time shift at another job and thought he should stop by the scene to see whether his help could aid the rescue. The twenty-four-year-old shuddered at the parental nightmare playing out on Tanner Drive—his wife having given birth to their second child just a month prior. Forbes headed into the backyard not knowing what to expect. As he walked up onto the scene, he did not realize he would be there for the duration of the rescue mission.

Perry, one of Boler's men, focused on punching through the rock, unable to take his mind off the little girl who rested painfully on the other side of five feet of rock-hard earth. Each time the weariness of the exhausting hammering crept into his hands, Perry thought about the end goal. The drilling shifts slipped by faster that way.

Firefighter Dave Holman, who worked at drilling in the rescue shaft late Wednesday night and into the next morning, experienced stints of confusion, frustration, and disorientation as he fought with the bit and hammered at the rock, chipping off pieces and spraying fine dust.

"Once you get down there, you lose all sense of direction. You really don't know where you are," Holman told *Austin American-Statesman* reporter Jim Phillips, who had made the five-hour drive to Midland to cover the rescue.

As the night had drifted silently from Wednesday to Thursday, drilling speed and the march diagonally did little to inspire optimism—though officials continued estimating a rescue time as bystanders clamored for updates on progress and results. A light sprinkling of rain gently fell. The dust, having become part of the air all around them, settled under the weight of the moisture. It only slightly muddied the area around the well. Southwestern Bell employees came in, installing a canopy to shield the well and rescuers around it from the elements.

Using jackhammers presented a not-so-efficient method to help advance Jessica's salvation. Although some of the world's most advanced drilling techniques put Midland at the forefront of the oil drilling world, there was no simple solution to the slow process of drilling through rock—sideways. Equipment designed to drill straight down uses its own weight to power itself into solid objects, pulverizing it into bits. Felice and the other drillers hoisted the jackhammer's drill bit into the wall, resting the remaining part of the hefty machine along their bodies—either their thighs or their chests. The gyrations shook their bodies, rattling their insides and rocking their skulls to the point of severe pain. If not handled properly, the jackhammer pumped the driller into the back wall of the rescue shaft. If the digger did not lean into the tool with a constant, full force, the pounding hammer bounced off the wall and into the digger's body.

Bentley went to an assistant fire chief again sometime in the wee hours of Thursday morning, feeling too groggy to even recognize which chief he had found in the midst. With the rock, they were dealing less with caves and more with mines. Though the rock limited worries about a cave-in, the idea nonetheless seeped into the minds of those hammering on the wall as dust fell and whole chunks chipped away with intermittent success. Their minds played tricks on them and led them to stray into their imaginations. Suddenly, it became clear to many of them that a total cave-in was possible.

"What really is needed here is somebody who works in a mine, because that is what you're dealing with. It's the same thing—hard rock—and they know how to get through that stuff," Bentley explained.

The notion put a thought in Roberts's head.

On Midland's outskirts sits a small international airport housing commercial passenger jets and nearby, smaller private aircraft are housed for other fliers. Oilmen also used a small municipal airport in North Midland for taking off and landing their small jets, allowing them quick movement around the state to various oil leases or to larger urban centers like Dallas, Houston, and Austin for business and political deals.

Pilot Aubrey Price does not remember the call coming in from his boss ordering him to prepare Clayton Williams's Sabreliner 40A for a quick flight. But he does remember the chain of events that led to a critical turning point in the rescue. The Sabreliner 40A was first developed by the military. The jet has a nose that sticks out like a duck bill, but it has a maximum speed of 480 nautical miles per hour and a range of more than 1,100 miles. It was one of a handful of aircraft owned by the local Midland oilman.

Since the early 1960s, Clayton Williams had made himself well known in the state. Most particularly, he had made his name throughout the Permian Basin for his oil and gas dealings. In the 1970s, his work in real estate started moving into ranching as he bought up historic cattle ranches in and around the Davis Mountains, near his hometown of Fort Stockton. Since then, Williams, known to friends, family, and Midlanders as "Claytie," moved around the state and country by air. In the early 2000s, notable motion pictures were shot on his ranches near Marfa, including *No Country for Old Men* and *There Will Be Blood*.

Williams was known for his audacious and aggressive maneuvering, which would prove critical in the race to save Jessica McClure. Though off-putting at times, his manner was deemed justifiable by some, and it was, perhaps, defensible as a key trait necessary to successfully navigate oil and gas, ranching, and real estate. Modern ranching could be contentious. As *Texas Monthly* writer Gary Cartwright explained in a February 1985 feature, the Williams ranches were breaking with decades of tradition, padlocking gates that gave throughway to other cowboys trying to reach their own ranch properties.

For West Texans, though, Williams had achieved what came close to celebrity status, due in part to how he exemplified the popular local belief that anything was possible with fortitude and hard work. The world would eventually be introduced to Williams during his 1990 run for governor; he would come to represent exactly what the public believed a Texan to be as he brashly defied the laws of politics in his own style of campaigning. The campaign for governor put a spotlight on Williams—and Midland—and on his brand of Texan. But several high-profile political gaffes ultimately doomed his move into political stardom, and he lost by 99,000 votes—about 2.5 percent—out of 3.8 million votes cast.

Following Bentley's discussion at the scene, highlighting a need for someone with mining experience, administrators at the MFD rooted around for the right set of hands. Williams, having made it known that he would make any of his resources available if needed, agreed to use his jet to ferry a mining expert to the scene. Several hours away, Dave Lilly, an inspector investigator with the US Department of Labor's Mine Safety and Health Administration, was already at work when he received a call from Midland asking for his help. Lilly's thin, tight frame could squeeze easily into and out of tight places. He spent his career doing it. The fifty-one-year-old mine investigator had moved to Carlsbad, New Mexico, in 1982. After receiving the call, Lilly made a call to his wife, Doris, at home.

"I need some clothes quick," Lilly explained, after which Doris started looking for a change of pants and a shirt. "I've got fifteen minutes to be at the airport. There's a girl trapped in a well in Midland."

For Doris, the call was concerning but not surprising. Her husband had worked in the mining business all his life—first in the coal mines of West Virginia, together with his brothers, in the 1960s. Lilly's colleagues—fellow inspectors on the fifty salt and potash mines in and around Carlsbad—had been involved in mine rescues before. Lilly stopped quickly by his house, picking up the bag his wife packed for him before leaving for the airport.

The pilot, Price, was already familiar with the rescue, having followed it as it was unfolding. He had worked in and around the rescue scene since late Wednesday, helping direct traffic and fulfilling various duties in his capacity as a reserve deputy for Midland County Sheriff Gary Painter. The extra work accompanied his full-time employment as Williams's personal pilot, regularly steering Williams's seven aircraft—various planes and helicopters—from West Texas to Houston and to all other points east and west.

Price called ahead to the air traffic control tower at Midland International Airport, knowing his assignment's success hinged on a dire need for speed. Cutting into the West Texas sky, bound for Carlsbad, New Mexico, Price needed all the advantages the Williams jet might afford—and then some.

CHAPTER 8

Price's radio call to air traffic controllers was direct. "Hey, guys, here's what I'm doing," he explained to flight controllers who could look out on the flat desolation of 360 degrees of open skyline obscured by little more than the Midland downtown buildings protruding strangely from the earth. Controllers took down his flight plan, which included a never-before-heard-of addition—"to help a baby trapped in a well."

Pulling strings is not so easy with air flight controllers, but Price needed to push the limits of regulations that would restrict his air speed under a certain altitude. Flying the 180 miles from Midland to Carlsbad would normally take thirty to forty minutes. Restrictions called for planes flying under ten thousand feet to stay under the 250 nautical miles per hour limit.

"Can you help me out?" Price asked.

Williams's Sabreliner 40A jet could easily max out its speed, getting him into Carlsbad faster than the usual flight time. Price wanted to peel away as many of those extra minutes as possible. With his request for a clear path and permission to fly at a speed not allowed at such a low altitude, the reply came back promptly.

"Absolutely," flight controllers affirmed.

Climbing at a high rate of speed, Price leveled out for only a few minutes before managing his descent. The crop duster pilot from Fort Stockton had come a long way. Williams first started using him to hoist him into the sky from Fort Stockton to Midland—an easy 107 miles.

The trip to Carlsbad took him about eighteen minutes.

"It was a time people actually wouldn't believe, because of the speed we were operating at," said Price.

Lilly waited on the tarmac—but not for long—when Price's jet appeared in the sky, skimming through the clouds in the east. Price barely rolled to a stop before Lilly climbed aboard to take a seat.

"We were taxiing off as he climbed on the airplane," Price said.

Moments later, they ascended into eastern New Mexico's baby-blue ocean of air, increasing their speed again to almost 500 nautical miles per hour.

Price called the Midland County Sheriff's Office dispatch, where Deputy Paul Hallmark manned phones, facilitating requests from rescuers and donations from local companies offering to do what they could to help save Baby Jessica. Much of Hallmark's work since mid-Wednesday had included hunting down equipment and organizing tools for delivery to the site.

"We've got Dave Lilly on board right now," Price called out to dispatch before taking off from Carlsbad, his sheriff's reserve radio still on and monitoring the rescue. "We should be due at Midland International within the next twenty minutes."

Lilly's feet touched Midland soil for only a brief moment before he found himself ushered into a waiting sheriff's vehicle and cruising the winding path from the airport to State Highway 80. The rescue efforts were approaching the twenty-four-hour mark, and the deputy's siren wailed. The drive from the airport to the Tanner Drive backyard would take longer than Price's jet flight from Carlsbad.

Robert O'Donnell finished his shift at MFD Station 2 around 7:30 a.m. on Thursday, October 15, before calling his wife to tell her he would be heading out to Tanner Drive.

"Do they need relief people out there at the scene?" O'Donnell had asked in a call to dispatchers after getting brief updates in bits and pieces throughout Wednesday and into Thursday morning. Firefighters on duty at other stations around town had picked up radio chatter that gave them occasional insights into the rescue operations and progress. Other stations were still busy responding to the normal daily emergencies. Running on a few of those emergency calls in the early hours of Thursday morning broke up the sleep O'Donnell would otherwise have sought. It is not as though he would have restfully snoozed through that night anyway. In moderation, he had fashioned a medication regimen to subdue debilitating migraine headaches that caused his excruciating insomnia. His difficulty sleeping made him the perfect kind of firefighter-paramedic—the kind who could barely sleep and shot straight out of his bunk when calls came through over a loudspeaker in the bunkhouse.

CHAPTER 8

O'Donnell's Thursday was supposed to have been spent with Chance, his young son. O'Donnell's wife, Robbie, counted on her husband not being home, though. She took Chance to her mother's house, automatically expecting him to go to the rescue scene as soon as he finished his shift.

O'Donnell's thoughts fixated on the idea of a baby in a well.

"I should be out there. I'm skinny, and they're going down holes, and I can fit down holes," he told Robbie.

O'Donnell drove to Tanner Drive, parking two blocks away and wading into the mass of vehicles clogging the street. News vans, volunteers' cars and trucks, police cars, and normal residential vehicles now lined every side street and alleyway. Available space filled up fast as onlookers, volunteers, and media flooded the area. The first satellite news truck to arrive on the scene that morning belonged to WFAA, based in Dallas–Fort Worth. As O'Donnell neared the rescue site, Dave Cassidy, a WFAA reporter, and a slew of other state and national media shot off live segments from the Sprague backyard. The humble yard looked like a movie studio's soundstage, full of cables, bright white lights, and camera equipment squeezed haphazardly into every pocket of space.

Cassidy, despite his satellite truck going ahead of him, almost missed his chance at the story completely. Rushing into Dallas Fort Worth International Airport, Cassidy carried with him enough clothing and toiletries for a few days away from home. Though the station had a reporter and crew on the scene, WFAA needed an experienced news reporter, and it looked like he was about to miss the flight. As airport personnel closed the door leading to his Midland-bound plane, Cassidy skidded to a stop, his small bag and a few notepads in tow.

"Wait! Wait! I'm with WFAA. I've got to get on that plane," he scrambled.

"I'm sorry, sir—," the airline gate attendant began.

"No, wait! I've got to get to Midland to do the story about the little girl in the well," he pleaded, somewhat out of breath.

It was the only time anyone ever reopened a gate for him. It has not happened since.

Cassidy was an experienced man on the microphone. He knew how to cover tragedy. He had been on the scene two years earlier when Delta Airlines Flight 191 came up short of the runway at DFW Airport. The crash killed eight crew members and 128 passengers. WFAA and Cassidy arrived just moments later.

There were obvious differences in the feeling of covering the race to rescue Jessica and that of the rescue of Flight 191's victims—though Cassidy had difficulty defining those differences. For hours during the night after the crash, WFAA had stayed at the scene reporting live during newscasts and staying on the air through noon the next day, though the coverage was not continuous.

"That was the biggest story I had ever covered," said Cassidy.

The difference between then and now was the amount of time the rescuers and bystanders now spent seeking a resolution. A certain relief could be felt in the finality of the Delta 191 disaster, where a rare weather phenomenon known as a microburst resulted in a powerful downdraft of air, pushing the plane down as it neared the runway. But with Jessica McClure, tension mounted, and the strain caused by the uncertainty continued to build as the rescue unfolded. "The longer we stayed there, to the end, it was the longest sustained tension of anything I ever covered," Cassidy said.

It was very different from the coverage at the Dallas–Fort Worth airport. There, ambulances sped to the scene, picked up surviving victims, and hustled them to the hospital. Within twelve hours, Cassidy's Delta 191 story was already moving from the airport to the hospital to victims' homes, developing follow-up stories, known as "day two coverage." Being away from the disaster at the airport minimized the emotional effects brought on by the initial visualization of the carnage. In addition, bystanders and reporters did not have close access to the grim scene.

Whereas horrifying acts of violence of unimaginable magnitude can overwhelm the most seasoned reporters, nothing is more able to suspend a person in stasis than a sustained exposure to the human element in crisis. Reporters and first responders tend to grow acclimated to the vision of a lifeless body at the scene of an accident or a crime. What stays with them is the emotion of the players involved and their response to what they are witnessing or managing, the subtle fleeting moments of a facial expression of a victim's family member as they react. By contrast, the constant rhythm of painful emotions at the scene of the Jessica McClure rescue sent pulsating waves of complex expressions of human struggle out into the world. As Cissy emerged from the home to check on digging progress or as drillers emerged from the rescue shaft covered in dirt, the world could see up close a visual human struggle played out on screen, over and over again. For hours, thousands watched as the rescue for Jessica pulsated and grew into a massive effort of universal concern.

Cassidy explained, "As the time went on and the fear [grew] that the child wouldn't make it, I had never experienced a sustained tension, nervousness, or dread covering any story in my life of that magnitude, because it was just so awful. You didn't want to get people's hopes up, because who knew? They were drilling into rock. Would the tunnel collapse? . . . It was just this group helplessness I had never experienced."

Glasscock and Beltran tried to keep themselves busy to pass the time. Reporters began keeping notes on the rescue's ongoing progress, marking hour seventeen, hour eighteen, hour nineteen. As many of the reporters, police, and drillers sought a refuge from Thursday morning's light rain and cold, they looked to the Sprague backyard and living room. Some spread out on the floor, others laid claim to a couch. One reporter recalled using a rock in the Spragues' backyard as a pillow. Most of the yard had been turned into a trampled mess by the early morning. No longer did a fence block the Spragues' backyard from the alleyway. Rescuers—after pulling off the gate—yanked the rest of the fence down soon after.

During the night, Glasscock started to decline. Feeling his body starting to shut down, he felt tired for the first time, realizing he had yet to take a break long enough to catch even a flicker of sleep. He made a collection of discarded pizza boxes, stacked them in a pile, and found them soft enough to use as a makeshift bed. It was just enough of a cushion for a few minutes of shut-eye; but even he had to laugh at the image of a pudgy cop passed out in a stack of pepperoni-stained cardboard.

"They might as well have been a pile of donut boxes," Glasscock chuckled.

As Glasscock rested, other rescuers welcomed Lilly as he arrived in the backyard. A flurry of attention swirled around him as he examined the operation and decided to reorganize drillers and adjust the pitch of the rescue shaft. Looking around, he must have wondered just what he had gotten himself into: from the giant green rathole rig to the maze of TV cable wires to worn-out oil field roughnecks. His experience in corralling small bands of rescuers kicked into high gear as they looked to him for direction. The cadre of drillers was ready to take orders and go to any length his command warranted. They had their man.

Lilly had drillers decrease their angle, aiming lower on the well shaft. "Other than that, you're just fine," he told the chiefs, affirming other aspects of the operation's course.

Though Lilly and the chiefs preached the word of an accurate, dead-on course to pierce the well shaft, Lilly later told *People* magazine he had to make adjustments to the trajectory.

"By the time I got there, they had already sunk a parallel shaft about twenty-nine feet into the ground thirty inches wide, and they were starting on a horizontal drift toward the well that would have brought them right into where the girl was," Lilly explained.

Lilly looked around him at the pandemonium of the scene. Though so many people constructed a massive rescue, the degree of volunteer involvement slowed the process down severely. "The police chief and fire chief were trying to organize the rescue, and it was total chaos," Lilly explained. Other firefighters agreed, having grown frustrated over the lack of control and—although good-hearted—the physically less-able drillers who continued into the hole through Thursday morning.

"If they would have let just the firefighters do it, I think we would have been done twelve hours sooner. They did not realize who they had. Some of us were in pretty good shape. We could have kept rotating in and out. We would have been fine," said Felice. "They got a lot of these good old boy expert drillers down there that were out of shape, and they'd run the drill down there for a little bit and then stop and run the drill a little bit and then stop."

These bravado-laced statements undersell the difficulty of the task. Operating a 40-pound chisel is equivalent to heaving a large iron Olympic weightlifting plate and holding it at waist level at about 90 degrees. Then, add in the ferocious energy of drilling, which means uncontrollable shaking. As the tip of the chisel presses against the rock, pressure builds in the tool's housing, mechanizing the jackhammer's rhythm. The resulting intensity of the vibrating bit is difficult to contain. The machine-gun rapidity of the chisel's tip firing into the rock requires a full engagement of one's entire strength. The equipment starts its penetrating vibration in the outer muscle tissues before seeping into the skeletal framework. Holding there for a moment, the vibration then works its way outward from the body's core to rattle every fiber and tendon. Lactic acid begins to build, first in the forearms then in the biceps, each burning with an intense fire. As the tool bounces in the driller's hands, the pressure they put on the device has to intensify and adjust so it can continue its chipping. It is not mindless work. Each maneuver and push into the rock requires an adjustment by the driller.

As they moved closer to Jessica, more adjustments meant a driller might have to wedge in sideways, letting the jackhammer chip away at an awkward angle. Then another adjustment. Again, and again.

CHAPTER 9

PAULA, GET YOUR GUN

Fifteen Years after the Rescue

WHEN A POLICE OFFICER IS SHOT DOWN, ATTACKED, OR otherwise targeted by the criminal element, law enforcement culture is emboldened, and when word spreads that a fellow officer has been targeted by a criminal enterprise, the brotherhood of law enforcement closes ranks quickly. After all, it is rare, making it a notable occurrence. According to the National Law Enforcement Officers Memorial Fund (NLEOMF), an average of fifty-four US law enforcement officials were shot and killed each year between 2010 and 2016. Most frequently, these deadly shootings occurred when responding to domestic disturbances. Despite this tragic outcome, the figure shows how infrequent police deaths can be despite the relative danger associated with law enforcement positions. According to the NLEOMF, officer fatalities from all causes in the line of duty totaled 243 in 2001. According to the FBI, there were about 705,000 sworn officers in 2001. Statistically, based on the total sworn officer population, there are many other things more threatening than a targeted shooting attack. A police officer who enjoys a round of golf is more likely to make a hole in one on a par three tee-off than he or she would be to get shot in the line of duty. And if that were not rare enough, a police shooting is only a few hundredths of a percentage less than the likelihood of getting hit by a meteor or asteroid.

So, the likelihood of a police officer getting shot in the line of duty is low. It has been especially rare in Midland. According to police memorial groups, three officers were killed while on duty in Midland between 1935 and 1961. One of those deaths was a traffic fatality, while the other two were by gunfire. An on-duty death did not happen again until 2019 when twenty-eight-year-old Midland police officer Nathan Heidelberg, responding to a home alarm system in an affluent area in North Midland, was shot in the chest by the homeowner.

The work of law enforcement exposes officers to society's most dangerous elements. Officers, whether on traffic patrol or working on technical details as investigators, interact daily with a criminal population. The danger in the work fosters a connection among comrades, and bullets do not have to fly to inspire a heightened sense of brotherhood on the force. In the summer of 2001, it quickly became clear that some measure of trust had been lost among the Midland police ranks when one of their own, Paula Bynum, was shot outside her home, and police officers seemed to be doing nothing about it.

March 21, 1999. The MPD's morning briefing was five minutes into its session before Officer Paula Bynum walked in the room. Red-eyed and less coherent than normal, she looked apathetic, weary, and to one officer, Sergeant Eddy Houchins, possibly stoned. Bynum, the police officer who had hustled Chip McClure to the scene of the rescue of his daughter, had done little to advance in the department's ranks. She was often named as part of the rescue team. She even got a small reference in a self-published book told from Chip's perspective. In it, Chip's writer notes his recollection of Bynum arriving in her police cruiser to take him from the paint crew job to the rescue site.
    Things had changed since then for Bynum.
    Sergeant Eddy Houchins was a veteran of the Midland police force. He had served since 1972. Sensing something amiss with Bynum, he quickly noted her appearance as she took a seat at the rear of the room. As Bynum's head bobbed—in and out of sleep—Houchins knew something was wrong with her.

Following the briefing, Houchins followed Bynum to Cowboy's, a restaurant in South Midland on Interstate 20, an area where much of the city's neglect continued through the late 1990s. The rundown core of the community had only minimal industrial businesses and even fewer retail establishments that would attract middle-class shoppers. Cowboy's became a hangout for cops in Midland, with many of them stopping by regularly for a coffee break or to grab a quick meal. The walls were covered in a brown wooden paneling hung with pictures of roughnecks and ranchers from long ago.

Sitting a short distance away, Houchins watched as Bynum fumbled with the breakfast menu.

"She sat down and immediately started trying to doze off," Houchins wrote in a report to MPD Lieutenant Clint Lovejoy. "When she opened her eyes, they squinted and [looked] blood shot."

Houchins watched Bynum gaze with a blank stare at her breakfast after the waitress delivered it to the table. Eyeing her watchfully, Houchins did not explain away the behavior, dismissing it as exhaustion or lack of sleep on Bynum's part; rather, he observed that Bynum was stoned on a narcotic. Houchins watched closely as Bynum clumsily attempted to open a plastic packet of jelly for two to three minutes. Finally, she reached for a silver butter knife, sliced into the package, and—retrieving its contents—spread it onto a piece of toast.

When Bynum received a call from police dispatch to report to the Metro Inn, a hotel located a few miles away, Houchins watched Bynum sit quietly at the table for five minutes "as if it didn't register."

Houchins called Lovejoy and requested a drug test for Bynum due to "reasonable suspicion." Lovejoy told Houchins to get Bynum and bring her back to the station so they could properly inform her of the accusation and need for the test.

As Houchins steered his patrol car into the Metro Inn parking lot, Bynum was leaving. Still dazed, Bynum drove past Houchins although he was waving her down from his police cruiser. He caught up to her several blocks away.

"We exited our vehicles, and I asked if everything was all right," Houchins said. "She stated that she was just exhausted, was having a lot of personal problems, and hadn't slept but about an hour."

"You're a liability," Houchins explained to Bynum before escorting her back to the police station.

A month and a half after Bynum took that drug test, which she officially passed around 3 p.m. that day, police responded to her home following an emergency call she made to dispatch from her police radio.

It was not the first time that police had noted a troubling trend with Bynum. And it was not the first time the department's administration took notice of major issues related to Bynum's behavioral health. In 1999, Bynum had come under additional scrutiny based on behavior reported by her friends and colleagues, but new leadership had been at the helm, as Chief Czech had retired in 1997, and a new police chief had taken over. By all appearances, the new leader, Chief John Urby, looked at the police force less as his own department and more like the municipal organization it was supposed to be. As Urby would hint, wayward officers would not fall under the protective cloak of a paternal influence. Urby himself kept a watchful eye on Bynum.

"The department doesn't belong to me, or the city council or the mayor. It belongs to the citizens of this city," the newly appointed chief told *Reporter-Telegram* writer Gary Shanks on Urby's first day, October 14, 1997. A tall and burly police veteran, Urby looked the part of a police chief. His hands were thick and muscular with a bear-like grip in a handshake. Despite his size, he possessed an approachable and friendly manner.

On March 21, 1999, Paula Bynum's friend Shala Baker drove to 4712 Country Club Road, where Bynum lived in a cozy home by herself. Bynum spent her weekends working outside on the lawn. She marched around the yard, towing bags of small clippings she pruned for the well-kept shrubs lining a walkway to her front porch. Although she kept a tidy house, the odor of stale tobacco lingered on the walls—trapped in the paint and in the furniture's cushions and wood grain. Baker arrived at the small home after having made a brief call to Bynum. Having known her for only two weeks, Baker, a twenty-six-year-old marathon runner, maintained that Bynum had sounded "odd" on the telephone. She had been worried about her new friend.

"When Ms. Baker arrived, she discovered that Officer Bynum was drunk and not making any sense," said Chief John Urby of the MPD in an internal affairs report on the incident. As Baker talked with Bynum, she realized she needed help in managing Bynum. Baker called Midland police officer Tabitha Stanley, alerting her to Bynum's shaky state.

Suddenly, Baker saw Bynum come from her bedroom carrying a pistol limply in her hand. Still waiting for Stanley to arrive, Baker followed Bynum into the garage, asking her to put the gun down. Struggling and concerned, Baker positioned herself behind Bynum, quickly grabbing her and still begging her to put the gun down. The two struggled over the gun, but Bynum pivoted and spun around. Bynum turned toward Baker and pointed the gun at her own head indicating an intent to harm herself.

The internal affairs report on the incident detailed the account.

"Officer Bynum kept talking about letting little kids burn up in fires," Urby wrote in the final internal affairs report. Later, Bynum would say the reference related to an incident for which she had been commended for a response to a fire where her actions had secured the safety of several children.

Finally, Stanley arrived to find Baker and Bynum wrestling over the weapon.

"What's going on?" she shouted in a panic. "Paula, what are you doing?"

Stanley put Bynum in a bear hug, begging Bynum to put the gun down. Bynum shook herself free and turned toward Baker and Stanley. Then she pointed the gun in the air and fired.

Baker and Stanley made another call for assistance to Midland police officer Laura Gilmore, who lived across the street from Bynum, hoping the additional assistance might bring Bynum under control. The three women successfully settled Bynum down and wrestled the weapon away from her; they then rushed Bynum to Midland Memorial Hospital. Although no one had been hit by the gun's bullet, Gilmore and Stanley called Sergeant Bernard Kraft to get more help. From the official reports, and according to police comments outside of the official reporting, a makeshift plot to cover for Bynum began to take shape. Firing a police-issued weapon required an official report, and even if it were not a police-issued firearm, failing to report the firearm's discharge would be a major problem for an officer if discovered by their superiors. Bynum asked for Sergeant Andy Glasscock's presence to help manage the situation, which had serious implications, given how her mental state would introduce new concerns about her ability to effectively act as a peace officer.

At the hospital, Gilmore approached Kraft.

"You don't need to say a word," Gilmore explained. "This will be handled through my supervisor."

Glasscock arrived and looked at the paperwork outlining Bynum's condition and justification for further inpatient mental health treatment. According to internal reports made by MPD and from statements by officers, Glasscock took the commitment papers and, as Bynum was getting ready to be carted away to Desert Springs treatment center, one of the only inpatient treatment facilities for behavioral health treatment in the area, he tore up the papers.

"This never happened," Glasscock said, ripping the paperwork into pieces.

In 2005, Glasscock would claim that he had signed the paperwork, though others stand by their accounts of his shredding the documents.

Ultimately, the incident was not handled through Bynum's supervisor, Sergeant Ken Rogers. Reports note, however, that Rogers was advised of the events as they happened. Glasscock had rushed to the hospital when he was notified, claiming later that he went as a friend—not as a police supervisor.

"This incident has brought a cloud of cover-up to the Midland Police Department," Chief Urby would later note in the internal affairs report. "This cloud could have been completely avoided had Sergeant Glasscock not abdicated his supervisory responsibilities."

On May 1, 1999—a month and a half after the incident in Bynum's home—Bynum received a pay raise following a review of her performance conducted by Sergeant Rogers, who said Bynum's performance exceeded expectations.

After Urby was told about the performance report on May 4, the six officers who knew about the incident, including Rogers, were reprimanded with a four-day suspension without pay. Rogers, knowing about the shooting and Bynum's tussle with fellow officers at her home, had still given Bynum a positive evaluation.

The whole affair was always going to be tricky to keep under wraps. Open records are difficult to hold under lock and key after an investigation is completed. Note of the incident in a performance review would not only risk exposing Bynum but also warrant questions about the broader police apparatus involved in the firearm discharge and hospitalization. Though not explicitly a cover-up, exposure of Bynum's mental state would have come under question had Rogers made a more open disclosure of her broader performance issues.

By Tuesday, May 4, Sergeant Tony Dickie sought out Lieutenant Jim Sevey after hearing about Bynum firing off the round from her weapon at her home. Sevey had not heard about the allegations, but he quickly initiated an investigation to look into the matter.

Glasscock sloughed off concern for the shooting.

"As far as the discharge of the firearm, I did not think much about it due to my concern for Paula. She was and is one of my best friends and this is why I called the hospital in the first place," Glasscock said.

Glasscock and the others laughed off the punishment. Officials allowed the officers to take their four days of suspension here and there rather than delivering it to each in one punishing blow.

A month later, Glasscock's pay was bumped from $2,965 a month to $3,113. At the same time, after findings in the internal investigation, Lieutenant Clint Lovejoy appraised Glasscock's performance at a 3.56 out of 5. Lovejoy gave him a 4-rating for decision-making and judgment. Despite the frequent issues and reports of brushes with questionable

CHAPTER 9

police operations, Glasscock and Bynum each pushed ahead and, while not ascending police ranks, still maintained their employment.

The neglect finally caught up to Bynum and the MPD on the night of August 3, 2001—more than two years after the first shooting in her garage. Police arriving on the scene found Bynum with two bloody bullet holes in the upper part of her left arm. Someone had targeted her, she said. And escaped.

A *Reporter-Telegram* first-year reporter covering the education beat filled in to report on the story in place of the regular police beat reporter, who had slipped out of town for a weekend jaunt across the state to sling beers with college friends.

Lieutenant Chris Cherry reached for his police-issued pager attached to the rim of his police trousers as he sipped on a cup of hot tea in downtown Midland's Ground Floor coffee shop.

"OFFICER BYNUM SHOT IN DRIVE-BY," the message read on the pager's screen.

Cherry called the dispatch operator at the central police station.

"Am I reading this right?" he asked the operator. Cherry was not the only one who reacted with disbelief when they heard the news about Bynum.

The news, however, was true—almost.

Bynum had been shot once in the upper part of her left arm, just below her shoulder. Police quickly swarmed Bynum's home.

Glasscock was the first to arrive. As he sauntered up the driveway, a woman sat in his patrol vehicle. She had been cruising the streets with Glasscock while he was on patrol.

Karen Carol, who worked as a fellow ER nurse with Lynne Glasscock at Midland Memorial Hospital, had grown fond of Andy Glasscock, and he of her—though the affair began as a friendship. Lynne caught on easily. The rift between the Glasscocks had widened, with arguments over pornography Lynne had found downloaded on the family computer. As the affair progressed, Glasscock even took Karen and his children on trips to his parents' lake house at Lake Nasworthy 100 miles away in San Angelo.

"Dad and Karen are good friends, aren't they?" Lynne recalled one of her children saying when the group returned from the lake.

That Carol was in Glasscock's police cruiser was wrong in a number of ways—not just because of the affair. Karen Carol also had a sexual relationship with Paula Bynum. MPD deputy chief Jerry Compton had become aware of the situation and had ordered Glasscock to never again have Carol with him in his patrol vehicle—a direct order handed down and intended to keep Bynum from losing control, according to police. None of this would come out until later.

Now, at the second shooting at the Bynum home, MPD officer Steven Bracken, who normally patrolled a trio of crime-ridden apartment complexes less than a mile from Bynum's home, quickly followed Glasscock on the scene.

Bynum, her hand patching the gunshot wound, sat with two bullet holes pouring blood. She recounted to the officers what happened that night, just after 11:45 p.m., rattling off the details. Two men in a "two-tone pickup" drove by her home, she said, and shot her while she was moving a water sprinkler in her front yard. They unloaded three rounds before driving away, slipping off into the night. The men made their escape too quickly for Bynum to get a license plate number, she claimed.

By the time Chief Urby arrived on the scene, Bynum's guns rested in Glasscock's hands. He had secured them, taking them away quickly and recalling his own trouble in 1999—when the previous shooting occurred at Bynum's home.

Approaching Urby at the scene, Glasscock did not intend to make the same mistake. "Here, I'm reporting it this time," he told the chief.

The next morning, a reporter from the *Reporter-Telegram* went to Bynum's home to find out what happened the night before and follow up on the story. A standard procedure, a follow-up news story is often more interesting than the breaking news story itself. Bynum walked the reporter through the incident. Wearing a sling and recounting the event, Bynum morosely outlined the path of the bullet.

CHAPTER 9

"It went through the front of the arm and came out the back, so I guess you'd call it a flesh wound. I'm in a lot of pain. . . . I don't like having to look over my shoulder," Bynum told the reporter. "It's highly probable that it had to do with somebody from the past."

Two houses down from Bynum, Jeff VonHolle, a two-year resident of Country Club Road, spoke with the reporter. VonHolle said he had moved his family to the area because of the comfort and convenience of a safe neighborhood.

"When we moved in, that was one of the most important things . . . the safety of the neighborhood," VonHolle said. "Living two doors down from a police officer, you would think that would make it even more safe." When VonHolle heard the shots, he ignored the popping sound of gunfire, thinking the sounds were probably just fireworks of some kind left over from the previous Fourth of July celebration. Even after the apparent drive-by, Bynum also agreed that her neighborhood was safe. After all, she would know.

Bynum said, "It's a very safe neighborhood. We don't have any problems with that neighborhood. I think somebody was just out for Paula Bynum. I think it's a very safe neighborhood."

With police cruisers on the scene, red and blue lights dancing in the normally quiet city block, investigators started combing the home for evidence.

Bynum described the two men as Black—a detail not released by police in any reports to the media. Across town, as officers flip-flopped on how to handle the investigation, nineteen-year-old Randel Tryon heard a knock on the door just a few hours following the shooting. The week before, Bynum had reported to Tryon's home in response to a domestic dispute. While attempting to resist arrest, Tryon had apparently kicked Bynum, bruising her liver and earning himself a charge of aggravated assault on a peace officer.

This time, police began to question Tryon at his home, in front of his family and in his words, "roughing him up." Police said it was a standard procedure—the way to investigate the crime properly. Chief Urby later said that the questioning of Tryon was not unusual, especially considering his previous altercation with Bynum. Tryon was supposed to have been the prime suspect, but police said they did not harm him in any way, nor did they take him into custody.

By Sunday afternoon, the regular crime reporter was back in the office at the *Midland Reporter-Telegram* and was listening intently as the story was reported. To devise a standard follow-up story on the drive-by, he began making calls to law enforcement contacts. The only thing to be culled for the record came from a few paragraphs of superfluous narrative in written reports and quotes.

"Currently, it is still under investigation," Chief Urby said. "We don't have anything to update at this point. The investigation still continues."

That was not enough for the reporter, who knew about the 1999 incident—as it was rumored, not reported. But the rumors worked out for the police department as well. Thanks to a couple of those poorly placed rumors, the chain of command was unable to fully put the officer-involved cover-up of the 1999 shooting entirely behind them. The reporter on the story had been able to get enough details about that earlier shooting to ask Chief Urby whether the current August incident could be an iteration of the previous shooting. The incident did not sit well with some in the department, and word started to seep to the public that Bynum's case was not clear-cut. The doubt had crept into the ranks of the police brotherhood, and despite Bynum's public contention that someone was out there targeting her, fellow officers were moored. There was no swirl of activity to seek out who might target one of their own. Despite the rarity of an officer being targetet—in the officer's hometown—there did not seem to be a rush to find the perpetrators.

Still, when questioned on this point with reference to the 1999 incident, Urby stuck with Bynum's version of the current event.

"In this particular case, that did not happen. Officer Bynum is the victim," Urby said. "That has nothing to do with the investigation. It's water under the bridge."

The news story did not gain traction thanks to the hush-hush of the police department. Their reaction was cause for concern for some of the news staff at the *Reporter-Telegram*. If the version of events touted

CHAPTER 9

by Bynum and the department was to be believed, the next news story most likely to emerge would be a follow-up on an arrest in the mysterious drive-by. In that scenario, a somewhat standard police report would outline how two men were arrested for allegedly shooting Bynum as part of some plot for revenge. Reports from internal police sources of other investigators probing the streets in an exhaustive search for the perpetrators would be expected. When cops get shot, everyone rallies.

But none of that happened, and an exhaustive investigation of the shooting and a resulting search for the suspects did not appear imminent. The nonchalance of Urby and the rest of the police department was unsettling considering one of their own had supposedly become the target of an unknown plot of revenge.

Word came in the middle of the week that something fishy surrounded the Bynum shooting. Tips began flowing in to the *Reporter-Telegram*. One call recommended that reporters should check into Bynum's past; another call maintained that news sources were not reporting the truth.

Reporters continued prodding Urby.

"There hasn't been anything changed since Saturday," Urby said sometime around Wednesday. "Officer Bynum is the victim. She is very respected on the force. She always does a great job."

Without much information to go on, reporters also jostled Tina Jauz, the public information officer for the city of Midland, who was doing double duty since Jim White, the information officer at the time of Jessica's rescue from the well, had been moved back to the patrol division several years earlier.

In a little downtown shop where only a handful of people would sometimes gather in small groups and chat, the *Reporter-Telegram*'s green crime reporter sipped lazily on a muddy cup of coffee and shook the occasional contact down for information or a new news lead. Across the room, the shop's owner chatted with a police officer. The reporter looked the officer up and down, waiting for him to look away so he could get a good look at his name printed on a gold plate just above his right breast pocket.

The owner of the shop himself had previously supplied the crime reporter with a good number of tips about Bynum's self-inflicted gunshot wound, having exchanged gossip and innuendo on his own. Now he sat with the officer and chatted openly.

The shop's owner waved the reporter over.

"Come over here," he quietly cajoled.

Without jumping into what the newspaper already knew about Bynum, the shop's owner began laying a light groundwork. The officer sat quietly and listened, and he initially appeared to be an altogether forthcoming source. The shop's owner and the officer tiptoed lightly around the Bynum issue, perhaps sizing up the risk of revealing themselves as sources, on the record, in a very public story about the shooting. At this point, enough could be cobbled together to suggest that Bynum had shot herself in a ridiculous grasp for attention. Her self-loathing boiled over the night of the shooting, knowing that her romantic interest, Karen Carol, was spending the evening with Glasscock in his car.

"All I need is a quotable source," the reporter said to the shop's owner and to whoever else was in range of the obvious inquiry. "I know that Bynum shot herself. Hell, all I need is someone to confirm it, and I can go ask the questions." As a Hearst Corporation newspaper, *Midland Reporter-Telegram* policy required named sources on the record for a story to go to print. Although the policy had been bypassed at other Hearst-owned enterprises, the *Reporter-Telegram* did not see itself as an enterprising gumshoe detective agency. Their bread and butter as a community newspaper meant covering local news in a less sensational way than the Hearst name might imply. There were still some ellipses around the sourcing issue.

The officer smiled without saying a word, and the reporter could finally read his nametag with a prefix set of initials indicating his mid-level rank in the department. The reporter gave up after a few minutes, when it was clear that the cop sitting there had nothing to add. When the shop's owner walked away, however, the cop leaned across the table and asked, "What is it you need?"

That's more like it, the reporter thought.

"What I need is a quotable source to confirm that Bynum did shoot herself," the reporter said. "If you can't give me that, then tell me where to look. Tell me the questions to ask. I'll ask the questions."

CHAPTER 9

The reporter's intention was to gather as many clearly articulated facts as possible and then pose a series of questions to city officials in interviews. For those who knew the truth of the investigation, their discomfort would grow with any sort of misdirection. And for officials like city council and city management, who might not be fully knowledgeable of the investigation's details, enough discomfort might develop to inspire their own questioning.

"Everybody in the department knew what was going on the whole time," the officer finally blurted as though the point had been bottled up. "Urby knew the whole time. Supervisors were stopping officers from making stops on vehicles that matched that description, because they knew that's not what likely happened. There's no way she should have a gun."

While the cop could hand over bits of information, the reporter agreed not to quote him directly. The officer had a family to support, and with a number of years before he could nail down retirement, he knew he would surely be fired for speaking so frankly with the media. The information could act only as a guide.

The reporter who had handled the initial Bynum story had gotten her hands on the phone number for Bynum's home. It was now Saturday, August 11, one week after the second shooting, and the green *Reporter-Telegram* crime reporter rubbed the piece of paper with Bynum's phone number in his hand, wondering whether he should call just for the interview. His skills as a reporter—not even three full months on the job—could be either a handicap or a tool. If he overstepped, he realized, he and his editors could brush it off as rookie overzealousness. But if he missed this story, then it confirmed his inner fear about the mediocrity of small market journalism.

The local news reporters keeping up with the story knew at this point that Bynum had shot herself, but validating this knowledge—so it could be printed and reported publicly—was still up in the air.

The *Reporter-Telegram* reporter dialed the Bynum phone number before he could think about what questions could be asked professionally without offending the woman. Under the limits of objectivity, he knew there was the off chance that Bynum had not really shot herself; but thoughts like that tended to kill the rabid hunt in a media world controlled by public-relations gurus and marketing whiz kids.

Paula answered the call. She had to be handled lightly, so the reporter dabbed and dipped, waiting to ask Bynum how she was shot.

"So, how's the investigation going?" the reporter bumbled. In part, this interaction would take the measure of his nerve. Did he have a strong enough elbow to wedge the interview back open when things had started to close up? A bold ability to stare down power, authority, and influence?

Bynum audibly stiffened, curtailing comments that held any kind of valuable detail. Her tone showed signs of becoming more fed up, and when her voice bordered on hostility, the reporter decided he had nothing to lose. Given the kid-gloves approach so far, he decided there might as well be some points on the board for him even if the game ended.

"Paula?" he asked, attempting assertiveness. "Did you shoot yourself?"

Another pause. It was not clear for a moment whether she would answer. The question hung in the air. The phone line might have gone dead, but he dared not interrupt her thoughts by asking whether she was still there and ruining the awkward silence. Instead, the silence mandated a response.

"What?" she asked, her tone monotone and tough.

"Yeah . . . did you shoot yourself?" he asked again.

This time without hesitation, she answered.

"No, I did not shoot myself," she said.

"All right then, thanks a bunch. That's all I needed to know," he said before hanging up the phone.

An hour went by before the telephone at the reporter's desk rang.

Paula's father was on the other end. He was not happy.

"Why do you go and drag up trash on people?" he asked, angry and fed up.

"When you hear it once, it's a rumor," the reporter explained. "But, when you hear the same thing repeatedly, you have to wonder." Still, he knew he did not have much to go on except for rumor and an unnamed source, who admittedly might have an ax to grind. Without something concrete in the form of an official statement or a named and publishable source, the *Reporter-Telegram* editors would not let even draft versions of the story be considered for print.

On Monday, August 13, two days after the reporter's direct questioning of Bynum, the city's public information officer released the "details" of the investigation. The same day, Bynum was placed on administrative leave with pay.

"The investigation concluded that Officer Bynum's gunshot wound was self-inflicted."

As the electrons from the neatly compressed press release transmitted line by line from the city's public information officer, Tina Jauz, onto the ancient fax machine paper at the *Reporter-Telegram*, the crime reporter who had doggedly attempted to exclusively uncover Bynum's shooting as self-inflicted had already moved on to another story. This time, he was interviewing residents at the trio of crime-ridden apartments where MPD officer Steven Bracken regularly patrolled, near the intersection of Wadley Avenue and Midland Drive. As he was questioning a number of residents at one of the complexes, the newspaper's photo editor called his cellular phone.

"Dude, you were right. We just got a press release from the PD [police department]," the photo editor said. "Paula shot herself." Even the photo desk editor was excited to have insight on the real story before it was made public.

Chief Urby made himself available for interviews, and he forthrightly started answering questions from the media. To his credit, Urby, without reluctance, responded to each question firmly and directly.

"I'm very disappointed that the investigation revealed the gunshot wound was self-inflicted," Urby said. "We just want to get this behind us and move on."

Urby knew this incident was not the first one, but at the time, the media and Midland citizens did not know the whole truth. Urby had no choice but to terminate her employment. He said he was waiting to make a decision on her fate as a Midland police officer.

"Where does the case go from here?" the crime reporter asked Urby, who said his officers were packaging the evidence regarding the case into a report for District Attorney Al Schorre, who would prepare charges of making a false report to police.

All was quiet regarding the Bynum shooting for a time—except, that is, for Bynum herself. The twenty-two-year veteran of the MPD knew nothing but the service she had provided Midland as an officer of the law. She had served as the first female member of the SWAT team. She was a member of the Honor Guard, and she was the first female canine patrol officer. She had hunted Chip McClure down to tell him his daughter had fallen twenty-two feet down a dank well shaft.

She knew nothing but being a police officer.

In the days following the affirmation of Bynum's self-inflicted gunshot wound—causing her to be placed on suspension with pay—she awaited her fate. Meanwhile, Urby, Jauz, and Rick Menchaca, Midland's city manager, would neither confirm nor deny Bynum's termination. They were sitting on the information.

On the morning of August 21, the district attorney met with several police officers, including Lieutenants Chris Cherry and Clint Lovejoy and Deputy Chief Jerry Compton.

The *Reporter-Telegram* had been asking questions about Bynum's resignation, and District Attorney Schorre did not even know whether she had been fired. He asked the officers whether Bynum was still on the force. He was answered with a raw, dead silence.

"Has she been fired?" Schorre asked the three officers, according to one of the men in the room.

No answer.

"The *Reporter-Telegram* doesn't know anything," Schorre said, passing off the idea of information about Bynum getting into the public's hands.

CHAPTER 9

"The MRT [*Midland Reporter-Telegram*] never knows anything," Compton countered.

Lovejoy retorted as Cherry sat silently, "Maybe they know more than we think they know."

Cherry held his silence in the room with the other men.

The reporter wanted to keep the story going. More had to be printed to justify the delay in revealing the truth to the public about Bynum's self-inflicted gunshot wound and whether there had been a deliberate cover-up. To him, taking a problematic officer's fake story and giving it purchase in the public should be more of an issue for city officials—and questioned. The *Reporter-Telegram*'s crime reporter went to the Ground Floor coffee shop to see whether his police contact was around.

The officer was inside, again sipping on a hot cup of tea, the tags held tight to the side of the cup with one finger.

"We can't talk here. Meet me after you get off from work around 10:45," he told the reporter.

The fear lodged in the mind of the reporter now. The reporter could not really know just what he was getting into. For all he knew, he had stumbled into someone whose interests were tangential to his own. The officer had asked to meet in a quiet parking lot of the sprawling grounds of the Permian Basin Petroleum Museum at night, just off the interstate. The area was isolated from public view. This could very well be the place where they corner, threaten, and intimidate nosey reporters. He decided to risk it anyway. The reporter told a handful of copy desk editors about his plan to meet the contact before he wrapped up the last of the night's reports around 10:45 p.m. He then pulled out of the *Reporter-Telegram* parking lot and made his way along Lamesa Highway, heading south toward the interstate to meet the police lieutenant.

At first, the only thing he could make out was the silhouette of the unlit museum building and several towering old oil derricks and other equipment on display. Across the museum grounds, a series of displays featured equipment from the years gone by in homage to the oil industry that had given this place a reason to exist. The cop was sitting in his white patrol vehicle and turned on his parking lights as the high beams of the reporter's pickup flashed over the facade of the closed museum. Pulling along the curving driveway of the sprawling building, it was not clear exactly what might be in store for the reporter in those next few moments. An occasional breeze pushed through the night air. The restored oil derricks on display throughout the museum grounds made ghostly eerie noises. In August in West Texas, the night air is still dense with dry heat even in the wee hours. Nothing could be heard out on this edge of town but the steady drone of 18-wheelers on Interstate 20 in the distance.

The officer—who required anonymity to protect his position of leadership and ongoing employment on the police force—had begun laying out a measured tirade against individuals in the police leadership who had sacrificed their own credibility and integrity to protect a flawed and privileged few. Revealing this information was not as concerning to him as getting caught as the source of the information. To a degree, he justified his actions, maintaining that pulling back the curtain was intended to protect the community. While not a direct threat, the behavior of police in mid-level leadership shirking rules and authority to secure their own reputations in the community had gotten out of hand. Whether the insight provided by the lieutenant could have resulted in his being fired is questionable. But revealing to an outside source the internal mechanics of an ongoing investigation as well as flawed internal decision-making by a police force would, at the very least, result in some discomfort back at headquarters.

The lieutenant was also not acting solely in the interest of the public. As the reporter would discover when he had culled enough information, there was a tender underbelly to the revelations the police officer provided. The lieutenant had become an outspoken internal critic of decisions made by city and police administrators on a series of policy issues. Chief among his complaints, however, was their decision to change the effective range for retirement from twenty years of service to twenty-five years of service. The policy change hurt him, and he had become vocal about it, which had earned him a less-than-ideal shift on night duty. Pulling overnight shifts limited his family time. He, indeed, had an ax to grind, and he had become disenchanted with the leadership's decision-making. For the lieutenant, this revelation to the reporter was his retaliation.

These kinds of whistleblowers are not new. They are as old as politics. It could be said that whistleblowers are more often the result of an outspoken critic losing an internal fight rather than simple do-gooders with a conscience.

As the lieutenant outlined to the reporter in the darkness on Midland's south side, Sergeant Andy Glasscock had continued disobeying the direct orders of his superiors—the same behavior that had, to an extent, led to Bynum shooting herself. For the police lieutenant, though, the department had become so accustomed to the strangeness surrounding Bynum and Glasscock that when the call came that she had been shot, there had been no rush to find the shooters. They knew very well that what was reported was a complete and total falsehood.

In the time between Bynum's placement on administrative suspension and her termination from the MPD, she was falling apart, and she did not try to hide it. On August 28, following the delayed suspension, Bynum was on the telephone with a friend who became concerned that Bynum might be a risk to herself.

Done hiding the truth, the MPD was compelled to release that information, clearly beginning to take steps that would distance themselves from Bynum's instability.

The real issue centered on why Bynum's termination was being delayed in the first place. And, as it would be revealed, the police administration's hands were full with other issues.

On Friday, August 17, a tip hit the *Reporter-Telegram* about a Midland police sergeant who was facing demotion or termination. It was Sergeant Andy Glasscock. He had been the first to respond to the scene the night Bynum shot herself. That night a woman, who worked with Glasscock's wife at Midland Memorial Hospital as a nurse, was in his vehicle. The woman, Karen Carol, was also, according to the police lieutenant, Bynum's girlfriend. Glasscock was seeing her as well.

This unfortunately was underlying Bynum's issue, and it was a key factor in what may have caused Bynum to shoot herself in the arm that night on August 3. Thoughts of Glasscock and Carol might have been what pushed Bynum into a jealous rage. It was revealed that moments before Bynum pressed a snub-nosed pistol against the fleshy part of her arm, she had spoken to Glasscock by telephone.

Deputy Chief Compton, knowing Bynum's sensitive condition, had ordered Glasscock to never have Carol in his patrol vehicle. However, the night Paula shot herself, Karen Carol was there—despite Compton's order and despite Glasscock's marriage.

Following a city council meeting on the morning of August 28, the *Reporter-Telegram* crime reporter asked Urby a series of questions pertaining to another story while waiting for a television news reporter to clear the area. The camera operator for the CBS affiliate Channel 7 continued to linger, looking on and listening in while taking his time to roll up a microphone wire. In these local markets, and in many larger metro media markets, the camera operators gather plenty of inside information by listening in. Many times, they gain access to police officers, investigators, and lawyers through a congenial relationship and are not seen as leak threats. They pass along information they overhear or generate off the record. Although still green, the *Reporter-Telegram* reporter knew enough that if overheard, the question he was asking might get passed along to other reporters back at the TV station. But unable to stall any longer, he finally relented and dove in to directly examine what Urby would reveal about Glasscock.

"Is there an officer about to be demoted from the Midland PD?" the reporter asked.

A smile spread in a slow stretch across Urby's face.

He began to lay out a cautious and benign answer, revealing nothing in the ambiguity.

"We do have an ongoing internal investigation. It does involve a sergeant. Hopefully, we get it wrapped up by the end of the week," Urby said.

"Does this relate to the Bynum shooting?" the reporter asked.

Urby said no.

CHAPTER 9

The reporter asked, "Did the incident that spurred the investigation come from the same night Paula Bynum shot herself?"

Urby would not answer directly but confirmed its lack of relation to Bynum shooting herself.

Urby looked back at the reporter with a cocky grin and said, "That about cover it?"

Glasscock at the time knew that a additional reprimand would land him in hot water once again. Having a woman in his vehicle was not a violation of police procedure if there was a waiver signed by the occupant—and, of course, as long as a deputy chief of police had not ordered that woman's absence from the patrol vehicle specifically.

Most members of the police department were aware of Bynum's affair with Carol. And they eventually found out about Glasscock's affair with Carol as well. Lynne Glasscock certainly knew. At least one officer noticed Carol on the scene at Bynum's home the night of the shooting. The only question that remained was if it could be proven openly.

The reporter continued asking department spokesperson Jauz to clarify when Urby would confirm details about corrective action he would take against Glasscock. And rather than ask directly about whether an examination of an officer's impropriety was imminent, the reporter decided to change tactics and assume that the internal decision was already down to a question of demotion or termination. There was a buzz of discontent in Jauz's voice as the prodding continued. Whether Jauz had direct knowledge of the police issue was not a concern because many public information officers are routinely left in the dark until the last minute. Still, putting pressure on the city to release information and putting it all on the record seemed to be the right course for ensuring accountability.

"I don't know who is telling you this," Jauz said to the reporter, finally coming to terms with what exactly was being alleged. "But I don't know what they have to gain by telling you that."

Just days after Jauz insisted that such a termination or demotion was not being considered, a flurry of electrons were once again transmitted line by line from city hall to the newspaper's fax machine. Each line relayed the skimpy facts surrounding the details of Glasscock's demotion from sergeant to patrol officer. It had become clear that Glasscock was indeed part of a dark cloud shrouding the police department operations and leadership.

The reporter paged Jauz. Then he called Urby.

Glasscock was once again in the spotlight. However, instead of playing unofficial spokesperson for the McClure family, he was facing a demotion and an embarrassing public divorce.

"I was learning more from the stories in the newspaper than I was from Andy," Lynne Glasscock said in March 2005. She was still married to Andy at the time. And more information began pouring into the newspaper from tips.

"The only way you can get this information on Bynum and Glasscock is to get access to their files," one internal police officer clarified to the reporter. "Everything they have ever done is in their files."

Filing a request for information obtainable under the Freedom of Information Act (FOIA) is a relatively simple procedure. However, if an investigation is ongoing or a prosecution has yet to fully process an acquittal or sentencing into a guilty verdict, the public's view of details related to any alleged criminal behavior will remain opaque. In the request presented to Chief Urby, the city council, and city administrative officials, the reporter specified the need for many items pertaining to Bynum's history with the force. Getting the issue fully into the light remained impossible without the documented records. Because of the restrictions disallowing *Reporter-Telegram* reporters from quoting anonymous tips and sources, extra effort went into trying to cajole one of the anonymous sources into finally naming themselves on the record. Otherwise, a set of documents, if they could be obtained, could justify making the details public.

While responding to a FOIA request is a relatively simple task, the MPD quickly pushed the request over to the city attorney's office, and in short order, it was passed along to First Assistant Attorney Chad Weaver. He quickly shot down every part of the request, maintaining each desired piece of information was not applicable under the open records laws due to technicalities contained in each document. To determine what should be made available under the watchful eye of the FOIA, Weaver informed the *Reporter-Telegram* staff that the entire box of requested files would be sent to Texas Attorney General John Cornyn's office for a ruling.

The refusal to submit information to the *Reporter-Telegram* was an obvious delay tactic, and it became apparent that getting the information on Bynum's past submitted to the public's scrutiny was going to be more trouble than expected. Fed up with the delayed release of the documents, the newspaper printed an editorial demanding answers. As the newspaper's history shows, the attention it calls to issues can make or break city politicians and administrators. But there are occasions when a newspaper's editorial page falls on deaf ears.

The day after the editorial ran, the *Reporter-Telegram* reporter visited Paula Bynum at her home, asking her to talk about whether she had been fired for shooting herself. Urby and Jauz—and Jauz still in the dark—would not say whether Bynum had been fired or suspended indefinitely. This visit followed the touchy phone call weeks before, during which the reporter had asked Bynum point-blank whether she shot herself.

Her giant police dog, Kody, peeked his large wet nose through the blinds as the reporter stood on the doorstep. Kody had served on the police force as Bynum's K9 patrol dog. Trained to take down fleeing criminals and would-be attackers, Kody was now enjoying retirement. From the same window, just a few feet above Kody's probing eyes came Bynum's round, bluish eyes. Weary and sad, Bynum looked at the reporter with her small lips puckered and stiff. The reporter gave a slight wave, trying to look as nonthreatening as possible.

"Ms. Bynum, I wanted to just ask you a couple quick questions. I won't take too much of your time. I promise," the reporter said.

She disappeared behind the thin window curtain, only to reappear at the door, just a second after he thought she was going to disappear within the confines of her home.

The door opened, and she came out. She was direct but not curt. And she spoke quickly to get this over with as soon as she could. It was clear that she wanted to close the door on this painful period of her life.

"I was released on Friday, but I actually resigned this morning," she said.

The reporter had just asked both Jauz and Urby that Monday whether the status of the Bynum situation had changed in any way. On the force for twenty-two years, she was three years away from receiving a pension from the department. Now, career lost, she knew it would be difficult to seek gainful employment. Bynum and the reporter did not chat much more after that. It was clear this was to be over and done with. But later that day, realizing more could be gleaned by encouraging Bynum to tell her story more fully, the reporter stopped back by Bynum's home to discuss further details. This time, she invited him inside.

"This is a friend, Kody," she explained to the dog. "He's not bad. He's a friend, Kody."

Bynum clearly had been proud to be a police officer. Framed portraits of her in uniform adorned the walls. On a living room wall, behind her couch, were plaques and certificates—a self-designed tribute to the impact she had made during her tenure as a Midland police officer. From time to time, she would glance up toward the wall from the couch, taking a drag from her cigarette.

When she went to a back bedroom to retrieve some up-to-date pictures of herself for the newspaper, none of which they ran, a knock on the door interrupted the silence in the living room. In walked another police officer, a friend of Bynum's who had stopped by to say hello.

Like Kody, the friend eyed the reporter cautiously, and Bynum came back into the room chatting without reserve as she thumbed through a large catalog of pictures. The friend never said anything to the reporter. The reporter thanked Bynum and, as they walked outside, he realized one last opportunity. For weeks, the truth around the shooting had been elusive, and it had been frustrating for him to deal with his editors and a set of corporate rules around anonymous sources.

"Paula, just one more thing. I almost forgot," he said as he casually stepped down the walkway of her front lawn.

She froze as if she were about to be asked to strip down.

"Can I see the gunshot wounds?" he asked.

She rolled up her sleeve, revealing the place where she had fired a bullet through her flesh. The wound had begun to heal, and a small scar like a cigarette burn marked her skin where the bullet had entered.

Later that afternoon, the reporter called Jauz as he began writing the story.

"Has Paula been fired, or has she offered her resignation at all?" he asked.

Jauz replied, "[Paula has been] on administrative suspension since August 13, and she remains in that capacity today."

Jauz had obviously been left out of the loop. Whether city officials were waiting to get their ducks in a row or merely choosing to let the details languish in the hope that their problem would just go away, it was clear to many that relying on the municipality to be transparent was a lost cause.

After the *Reporter-Telegram* editorial ran, Urby had obviously been shocked. A local news channel ran video footage of the paper editorial, which sharply asked Urby to come forth and spill the beans. Be transparent. Stop protecting wayward officers. Ensure the citizens that proper procedure was followed.

Stunned by the editorial, Urby called for a meeting with Meta Minton, the editor of the *Midland Reporter-Telegram*. Little did Urby know he would get a meeting with a broader array of staff, including publisher Charlie Spence, city editor Gary Ott, and the reporter who worked his way through the Bynum story.

They all gathered together so Urby could discuss the issues.

Urby walked in like a bear. His giant paw-like hands engulfed the reporters' in a hearty shake. He charismatically engaged the group, but not without a tinge of nervousness.

Minton and Ott exchanged greetings. The cheesy facade clouded the importance of the situation. Urby began to explain, but not before Minton reminded him that the meeting was on the record. Urby paused. He looked up, perhaps realizing the trap. If he stopped talking, it would appear as though he were hiding information regarding

Glasscock and his possible role in what might have caused Bynum to shoot herself.

"Sometimes, it's very difficult to air out dirty laundry, because it's very embarrassing, not only for the police department but also for the city," Urby said. "When those details come up during the course of an investigation, I find it very difficult to come up with the blow-by-blow details."

The newspaper staff knew Urby had been aware that Bynum had shot herself early on. To the degree that he could, Urby relented to confirm that point, and he provided the confirmation everyone had been looking for.

"I already had doubts [that night]," Urby said. "In the sense of protecting our people and their confidentiality, the investigation was unfolding."

That was a reasonable admission at this point. But it was uncomfortable to have the police chief in your boss's office and not have him do some major housecleaning on all the details.

"What would compel your department to order Glasscock not to have [Carol] in his patrol vehicle?" the reporter asked.

Urby started to answer, but he quickly diverted his attention to Minton.

"You see? These are the types of questions I feel uncomfortable answering," Urby said.

CHAPTER 10

GOOD GAWD!

Oil Country, Spirituality, and a Faith in Survival

THE SPRAGUES, PETRONELLAS, AND OTHER FAMILIES had been planted in West Texas on Tanner Drive for some time. Midland, like several cities across the country, consider their placement to be in the middle of the "Bible Belt," the term of endearment or derision, depending on the perspective, for an area across the southern United States whose inhabitants are perceived to hold an uncritical belief in the literal accuracy of the Bible. Others proudly claimed to be the "buckle" of the Bible Belt and considered the term a proxy for the overall gravitation toward a more general biblical faith.

When Dutch Lunsford finalized the deed in December 1965, he established his family at 3407 Tanner Drive just a few doors down the street from the Moores' house, where Jessica now was helplessly wedged in the well. For an oil field carpenter, moving into the upper-middle-class Permian Estates meant better education for his remaining children and a better neighborhood compared to the ailing one he left on South Midland's dusty streets.

Although they had worked their way up into the middle class, people in this area still knew strain and struggle. It was never far from their minds. Some experienced it through the travails of growing up on a Depression-era farm. Others knew it well from parents who had abandoned them. People like Dutch and Margie found much about their move into this relatively successful class a surprise. Men like Dutch had transplanted themselves to Midland twenty years earlier, when World

War II drove US Army recruitment efforts at full force, churning out new, young soldiers. Midland produced thousands of young men, who left their homes first for basic training before progressing to more specialized training to support the war effort. Midland Airfield became a training ground for thousands of these young men who had been selected to become bombardiers. The airfield was launched in 1942 on a 240-acre tract that had been scouted six months before the Japanese bombing of Pearl Harbor. Army Air Corps officials approved the site, which had previously been a flight school in the 1920s, and they eventually identified twenty-three bomb ranges within a fifty-mile radius of the field. In January 1942, a month after the US entered the war, officials set the location as an exclusive training ground for bombardiers to learn how to use a bombsight device created by Carl Norden for the US Navy. In sixteen-hour days, new soldiers poured into West Texas from all over the United States—as well as from other allied nations—crouching in Beechcraft AT-11 planes and learning how to rain hell. Dutch was among these young men.

Before Dutch could graduate from the bombardier school, however, Hitler's reign sputtered out, and Japan's military collapsed. The war over, Dutch stayed behind to build a career and family in the oil field.

"Round 'em up, boys!" Dutch called out to his gentle band of roving Boy Scouts, pleading for tiny slips of paper.

The scouts in his son's troop met Monday nights in a tiny Quonset hut taken from an old barracks at the Midland Airfield—which was slowly shifting to become a commercial airport, the precursor to what would eventually become the home of Midland International Airport. One by one, the boys were called to the front to place their sheets of paper into a hat for a shuffle. Each February, the celebration of Scout Sunday meant that the troop would visit a church chosen by way of consensus. Many times, a consensus could be manipulated for the greater good of a scout's rearing—or that is what Dutch's youngest son, Doug, figured after he ambled up to the hat and peeked in to see the cast votes.

When the votes were tallied, Midland's Free Will Baptist Church came out the winner, even though most of the scouts regularly attended the West Side Baptist Church. When Doug had peered into the vote-collection hat, he saw many more scraps of paper than there were boys, and he astutely determined the scoutmaster had pulled a fast one on the boys—perhaps, he later surmised, in an effort to get in good with a potential employer who attended the Free Will Baptist Church.

Many a man had been saved in some way thanks to a church in West Texas. For the greater good of a man's spirit, roots and community emerged from these churches, and this could not have been truer for Dutch. However, the church community also played a role in the greater good of the good man's pocketbook. In a town like Midland, a field hand, engineer, or banker never knew when he would stumble into a business deal or into a manager looking for a good man of God—one who knew how to work and advance the interests and needs of a company. So, when Dutch had a line on a job from a field manager who attended Free Will Baptist, he used his sleight of hand with the scout troop vote.

The transition to church membership by the Missouri farm boy, who knew a good can of Falstaff beer when he saw one, was not an overnight shift. The tug of the spirit came in the form of Eugene "Gene" Zoellers, a Navy veteran with slicked back hair greased down tight, pockmarks like the surface of the moon on his face, and a knack for the harmonica. Zoellers had an eye on Dutch, stopping by the house on Sunday nights with a group of other men from the church. Zoellers's wife, Barbara—known for her bright red hair—went with the women to a group of other assigned homes to preach the word of the Lord.

Dutch met them for the first time a few weeks before the Scout Sunday vote.

To some, Gene Zoellers might have come off like a snake oil salesman. To the optimistic, though, Zoellers's presence was as regular and required as the milkman. Preachers like Zoellers moved around, convincing each of their dusty oil field brethren of the threat of alcohol's eventual path to debauchery and frivolity.

CHAPTER 10

When Zoellers made his own visits to Dutch, the oil field carpenter answered Zoellers's knock rapping kindly on the glass storm door. Dutch's dingy work clothes still hung loosely on his wiry frame.

"Sure. Come on in, preacher," Dutch welcomed. "Can I get you a beer?"

The visits to the house continued for a few weeks until the preacher would bend the stubborn will of the Missouri farm boy. Soon after, Dutch brought his scouts to the Free Will Baptist Church and began to attend as regularly as a member of the church choir.

In a place where cash sprang from the soil, miracles seemed to be a daily occurrence. The boom-and-bust swing known by Dutch and several of the others who lived on Tanner Drive taught them to understand the hard times, and the church toed the line when the economy slowed. Over time, Dutch and his wife would take in other people's children, even adopting them in one or two cases. No one had their own bedroom. Kids moved in and out regularly. Other kids would find comfort in a bed there with regular meals. Even after their childhood years, some would regularly stop in to visit with Dutch and Margie, in several cases recognizing them as the only parents they felt they ever truly had.

By the time Tanner Drive was swarmed with visitors from media to rescue volunteers, many of the residents knew how to handle unexpected guests. And over the next few days, they would have many strangers sleeping on their couches, using their showers, and tying up their phone lines—especially after El Paso's Associated Press reporter Holden Lewis ran news across the wire about Jessica.

Lewis had been unimpressed with the rescue story when it first emerged. The Associated Press office in Dallas called asking him to make some calls to see just what was going on in Midland.

"It's a little girl," the editor said, according to Lewis. "The story has a lot of pathos."

Like the *Midland Reporter-Telegram*'s Crimmins, Lewis thought of a wishing well, picturing a four-foot-wide brick-lined wellhead and easily descendible with a firefighter and a ladder. Phone it in, he thought. Soon, though, he found himself on a two-hour flight headed east from El Paso, bound for Midland. In a duffel bag, he had quickly stuffed a few items along with a flashlight and a supply of batteries.

The story he sent out on the wire turned heads around the state and left reporters and editors watching for updates. Matt Quinn was watching in WFAA's newsroom in Fort Worth—the same newsroom Abraham Zapruder had waltzed into for an interview, holding fresh footage of John F. Kennedy getting shot by an assassin in 1963. WFAA had carried Texas's news profession into the twentieth century when, in 1922, the station began broadcasting from Dallas over 150 radio watts. The station eventually became the first Class B station south of St. Louis.

While the station nailed down history for Texas radio, it held power nationally in television because CNN and ABC used it for content when news broke in Texas.

Now, Quinn read the words as they spilled over the wire and looked at reporter Dave Cassidy.

"What if we have one here like that in our own backyard?" he asked.

The WFAA news truck left Dallas–Fort Worth Thursday morning, headed for Midland. It was the first sign of growing attention from beyond West Texas.

The story of Jessica's rescue was making its way into regional news outlets with print reporters carrying coverage that raised only a few eyebrows. As the story spread beyond the region, however, calls started clogging emergency dispatch centers in Midland. "Doctor Jack" sat at a terminal at Midland's Central Fire Station taking calls. Like many of the emergency response workers in Midland, Jack Williams, a firefighter-paramedic for the MFD, logged several hundred of the thousands of calls that would eventually stream in from around the world.

His nickname was not a term of endearment. Williams went into serious detail when reporting patients' conditions as he rolled them from the ambulance into emergency rooms. Williams bypassed ER nurses, asking for physicians to relay information about "his" patients. He was frank. He was to the point. And he took his role in emergency health care seriously.

CHAPTER 10

Soon, the calls began pouring into the call center. Callers with rescue ideas ticked off their innovative solutions. Others offered resources or their services as volunteers. Cindy Green sat near Doctor Jack as he took the calls. Cindy was a regular dispatcher for the department, and three other firemen volunteered to field inquiries when they began overwhelming what Cindy could address. Call-takers had been told how to field the onslaught of calls: Thank the caller profusely. Write down their contact information. Take their message. Tell them the message is being sent to the authorities.

"Y'all get that little girl out of the well yet?" a caller asked Williams, who answered a dispatched call to the 911 line.

Williams replied, "No, sir, we haven't. We're still working."

The man on the other end began to outline his own plan to rescue Jessica.

"Well, I can tell you how to get her out. Go down to the hardware store and get yourself some PVC pipe. Go get you some gauze, tie it on the end of it. And take some superglue and saturate the gauze. Put it on her head, and pull her up," said the man.

"The message is being sent, sir," Williams explained.

Others had plans—both better and worse.

While some wanted to send their kids down the well to pluck Jessica from her likely tomb, one man with a pet monkey called through to Williams.

"Now my monkey has got a harness," the man explained. "The monkey will grab hold of that little girl and pull her up."

As Williams fielded his calls and rescuers continued their dig toward Jessica, sixty-two-year-old Dorothy M. Chandler sat in her home in Endicott, New York, watching news coverage of the rescue effort. Chandler's plan called for an organized union of Little People to respond to Jessica-type situations. When required, they could be dispatched all over the country, she concluded.

"All over the United States there are Little People or what used to be called Midgets," she typed in a letter to Robert O'Donnell a year after the rescue. "They even have an organization called, 'The Little People of America.' Many of the members are only 18 to 20 inches tall. Some might even be smaller than Jessica, but the difference is they are adults and could understand if given directions and once lowered down there could talk to her, ease her fears, give her a toy, see if she was hurt."

As Dorothy watched the news coverage with her husband, Floyd, she hoped to see someone had the foresight and genius to implement her idea. "Oh, they will think of it soon," she said with excitement over the ongoing rescue efforts.

As reporters started to pounce on the story, the floodgate of callers opened even wider. Dispatcher Cindy Green put her hand over the receiver of the telephone in shock. Her face formed into a horrified concern.

"This is the Canadian Air Force," Green whispered to the other call-takers. "They have a demolition team on a helicopter that's in the air. They're flying here to explode the rock in the tunnel. What do I tell him?"

Williams smarted off, "Well, tell 'em NORAD has them in their sights, and they're gonna shoot them out of the sky."

"I'm not going to tell them that," said Green.

"Well, give me phone. I'll do it," Williams snorted.

CHAPTER 11

THE LAST HOURS OF ROBERT O'DONNELL

O'DONNELL'S BRAND-NEW PICKUP ALSO SERVED AS A sign of his new lease on life—one he started in Lubbock. He had even put a follow-up call in to the dealership in Midland from his room at the Koko Inn when he finished work on Friday, April 21, 1995. When he bought the truck, he never imagined his last drive would be across his stepfather's desert ranch.

Isolated out at the ranch in the rural countryside, O'Donnell had spent the weekend watching the ongoing news coverage highlighting the rescue attempts in the aftermath of the Oklahoma City bombing.

"He didn't like being out here on this ranch," his brother, Ricky, reflected. "He didn't like being in the country. He liked being in town around people. He didn't know his way around out here. There were only a few places he knew how to get to."

"Robert was a perfectionist. Everything he did, it didn't matter what he did, he did good—like dressing," Ricky described his brother. "When he left the house, he looked as good as if he stepped out of a book somewhere."

When he fled the house in the middle of the night, Robert wore a pair of freshly starched blue jeans his mother had pressed just hours earlier. He wore them over a pair of gray jockey shorts. On one foot, a white cotton sock. On the other, a blue argyle sock. He had thrown on a purple, red, and green sweatshirt, and he had hurriedly slipped on a pair of brand-new boots custom made by the Franklin Boot Company in San Angelo, Texas.

CHAPTER 11

As he slipped quietly out the front door—its paint flaking from years of desert wind—he was careful not to let the boot heels click hard on the wooden front porch. Before hustling away, though, he had clarified his intent on several scraps of paper. Scribbled in large black print are the words:

No help from no one but family.

On another:

You all are my reason for living. I am just tired of this world.
Love Dad.
PS: Don't ever forget me. Don't blame anyone but me. REO.

And another:

Casey I love you so much and Chance you're just like me. Love you both. REO

Yvonne, hearing the clothes dryer buzz, awoke to find O'Donnell's bedroom light on and the closet door open—the same closet in which the .410 shotgun was stored. By then, however, O'Donnell's truck had traversed the seldom-traveled back roads of the ranch. For more than two hours, he drove. In the dead of night he found a certain familiar loneliness. The same loneliness had enveloped him when the spotlight dimmed, and no one cared anymore that he had saved a little girl's life. Privately, he rejected the notion that salvation was the result of his own effort. He often looked at the photo album his mother-in-law had put together for him.

"I ain't no hero. We all did that. Get rid of that damn book," he had said, frustrated over his work environment when he was still fighting fires and plugging IVs into patients. The frustration had grown into anger. He reached down with a single hand and picked up the album. It smashed against the wall, making a furious flurry of pages ripple and spray torn from its hinge. Today the album rests with the same flimsy hinge—broken by the throw—in the corner of a top shelf of his ex-wife's closet.

"You're our hero," his mother-in-law had argued with him, and still would, despite her anger at him.

But all the while Robert O'Donnell was building extensive résumés with a description of the Jessica rescue. Other notes found among his belongings—mixed in with those taken during the days of fighting over the movie about the struggles of trying to rescue Jessica McClure—included names and addresses of people he wanted to hit up for jobs. The notoriety was not enough, it seemed.

O'Donnell's only relief came from his black duffel bag. Retreating to its insides in those early morning hours on April 24, O'Donnell tugged on a few of the bottles' caps, pulling out a few capsules.

Steering through the wilderness, he edged suddenly away from the faint dug-in dirt tracks of the makeshift road, veering off toward a grove of tall mesquite, their thorns hidden among the budding green leaves of the new spring season. He bounced up and over the bushes, coming to a stop as the greenery flipped back up behind his pickup. With the trees engulfing the truck, O'Donnell was hidden from view. From the road, O'Donnell's vehicle could not be seen—camouflaged in the spray of desert bloom.

CHAPTER 12

DAY TWO NEARS END WITH HOPE FOR A BREAKTHROUGH

IT WAS APRIL 1987, AND O'DONNELL'S SONS, CASEY AND Chance, were dancing around their grandparents' backyard, on the hunt for Easter eggs.

Holidays like Easter usually called for at least a small family celebration with get-togethers, backyard barbecues, and dinners. And as they had on many occasions before, Robert and Robbie loaded their two boys up for family dinner at the home of Hartwell and Joanne Martin, Robbie's parents, located in the heart of Midland on Princeton Avenue. The Martins had bought their quaint but spacious house in April of 1968. It served their family well, enveloped in the shade of large oak trees in a safe community.

The quiet and deliberate Hartwell—a junior high school football coach—and his elegant wife, Joanne, continued cultivating their family after the kids were grown and living on their own. The home was filled with grandchildren, giggles, and occasional roughhousing.

On this Easter in 1987, just six months before Jessica's perilous struggle, O'Donnell looked on, pacing around the backyard with his hands clasped behind his back as though he were inspecting a scene. The boys bounced and plucked eggs from their hiding places, placing the plastic, hollow shells in small baskets near Robbie on a picnic table. Casey, almost ready to give up his pursuit at one point, caught his father's attention.

CHAPTER 12

"Look at that one way up there, Casey," O'Donnell said, his long arm pointing about ten feet high in the crook of a tree limb. "You see it?"

Casey moaned back, "No."

"You want me to crawl up there and get it?" O'Donnell asked, already inspecting the tree's trunk to determine the best spot for a foothold. O'Donnell reached two stringy arms up high, grasping a branch tight with both hands. His height stretched well above the first branching crook of limbs. O'Donnell grunted a little and pulled hard, propping himself up into the crook of the tree.

"Oh, Robert!" Robbie called out playfully. "Robert, what are you doing? Casey, how did he get all the way up there?"

The group giggled, Robbie alongside her mother as Hartwell recorded the holiday with the family's video camcorder. Their playful laughter grew.

"Call the fire department to get Robert down. He's going to fall. Call the paramedics," someone else called out, laughing. "They're gonna have to come save you."

The family banter had gone on like this from the moment Robbie met O'Donnell at the urging of her sister. Robbie was the blonde Midland High School cheerleader and O'Donnell the fun-loving older boy who raced motorcycles. You had to be strong to race the heavier class of motorcycles, as O'Donnell did.

He attempted the military, but a foot fracture effectively cut off that avenue. Instead, he hung around Midland, enrolling in classes at Midland College to get a diploma via a Grade Equivalency Exam to make up for dropping out of Midland High School in the tenth grade.

"He was fun. We just really had a good time," Robbie said. He had his own apartment. He had a job. The money afforded them fun times out at the movies or dinners—things her friends' boyfriends could not provide.

"And he was just so funny," she said.

Soon, it was just time to get married with no proposal fancy enough to remember. At two o'clock in the afternoon on December 2, 1978, the two married, thus fusing the lives of two people from different worlds.

DAY TWO NEARS END WITH HOPE FOR A BREAKTHROUGH

On February 16, 1979, O'Donnell started working with the MFD, still developing his skills in the field. By 1980, O'Donnell had found his niche, enrolling in emergency care classes before studying firefighting at Midland College. He moved on to other advanced paramedic training courses. In his fifty hours of classes, O'Donnell never scored below a 3.0 GPA, occasionally nabbing a 4.0.

None of the criteria listed on his résumé played in the decision to send the fun-loving firefighter down the rescue shaft and across the tunnel to pull Jessica out of the well. The decisions had already been made at the scene long before Friday morning when O'Donnell would be the paramedic to reach for her. In the end, it was that lanky, lean body that made him capable of doing the job no one else could.

October. More decisions were being made at 4:30 a.m. on Friday morning as momentum grew and mine expert Lilly crawled deep into the rescue tunnel, pounding where the driller had plunged a series of round holes so he could chip away the rock in larger masses. As Czech and Roberts stood above, they had decided a paramedic would pull the baby from the shaft as soon as they broke through. Getting an assessment of the girl's condition as soon as possible could be done only by a paramedic or a doctor. Reaching her in such tight quarters would be difficult anyway.

O'Donnell, who had been standing by all day Thursday, became their man, with firefighter Steve Forbes standing at the ready for O'Donnell to pass the girl along to him. Work remained, though, and through the dark hours of Thursday morning, many were skeptical whether progress was being made. O'Donnell called his wife at work later in the morning to tell her the news that he had been selected to reach for Jessica inside the shaft.

"Oh, are you gonna be a hero?" she playfully flirted.

O'Donnell grinned with a slight laugh. "Whatever; you're being silly," he retorted.

Rather than wait around for rescuers to fumble in the effort to save Jessica, Bill Jones knew his idea to save the little girl would work, so he decided to make it happen with or without permission. The Midland restaurant owner—described in the media as a "Midland entrepreneur"—heard about the power of the hydro drill operated by ADMAC, Inc., in Houston.

"Hopefully, when the new equipment gets in, we'll go to work with it then," he told reporters at the scene around 2 a.m. on Friday.

Although some of the ideas to save Jessica were phoned in by Good Samaritans hoping to help, others like Jones and Ronald Short rushed to the scene, insistent on their idea's viability. The five-foot, four-inch Midland roofer, Short, stood before Czech and Roberts offering to slip down the well shaft, the one Jessica was in.

"Well, you'll never fit," it was argued.

Suddenly, Short collapsed his shoulders, pushing them to meet each other near his chin. Born without a set of collarbones, Short was able to bring his shoulders in to meet at his chest, leaving his girth to measure just fifteen inches wide.

They put his proposal in the queue with other outlandish potential ideas, along with that of Jones when he arrived on the scene preaching the gospel of the hydro drill. Each was told that their ideas were among the top considerations.

The 8,000-pound machine delivered by Jones powered water through a long metal gun like one found at a neighborhood car wash. This cannon, though, pumped water through rock with 35,000 pounds of pressure. Getting the machine to the scene would be a problem.

In his 1997 book *Halo above the City*, Chip McClure claims that he and Jones had coordinated the transport of the hydro drill, urging the chiefs to make it a priority. In a Facebook post in 2013, Chip explained his experience at the time.

"I have been awake for almost all of the last forty-eight hours and have eaten very little, the feeling in the pit of my stomach keeps me from any form of real rest and food has little appeal," he said. The frustration and fear were building, and much of his time and attention were committed to helping Jones get the hydro drill to the rescue site. Chip said he went outside at some point that night, recognizing the desperation of the rescuers now going into the well shaft and the slow progress being made to cut through the rock.

"If they can't figure out a new technique she dies. If it rains, she dies. If we don't get her out soon, she dies," Chip said in his Facebook posts about his thoughts at the time.

As the world narrowed its attention onto the scene long enough to catch a glimpse of Jones being interviewed on the news, those on the other side of the globe would see confirmed all their preconceived ideas about the people of Texas. Jones was a gallant-looking bold figure, heavy-set with the kind of Texas twang one would expect from such a man. His white straw hat perched high on his head, the tips of his fingers tucked gingerly in the tops of his pants pockets.

"They'll be able to drill a foot an hour," Jones assured rescuers, informing them of the drill's power. The enticing idea drew a number of concerns besides the one of whether Jones was just another nutcase. Still, considering their position of drilling around an inch per hour, it had to be an option.

After going first to Dallas to the company Jones's sister-in-law worked for, officials there led the fifty-four-year-old Jones to Houston to pick up the desired equipment. Getting the equipment to Midland was another problem altogether. The massive machine was on the other side of the state. Jones later told longtime *Reporter-Telegram* staffer Ed Todd of his call to the US Air Force in San Antonio, requesting the use of a Lockheed C-5A Galaxy "or something like that" from Kelly Air Force Base.

"What rank are you?" Jones claimed the Air Force major in the Flight Operations Center requested.

Jones replied, "Civilian."

"I'll get back with you," the major replied.

Federal Express, instead, went into action, arranging for the machine's transport to Midland International Airport. "It was just as simple as asking," Jones told Todd.

In the early morning hours of Friday, the Federal Express jet landed in Midland as coordinators then tried to figure out how to get the machine

CHAPTER 12

to the Tanner Drive backyard along with a 10,000-gallon water tank the city of Midland owned.

The reports from Jones about the potential use of the hydro drill fueled reporters as they fumbled for fresh angles on the rescue. Indeed, by Thursday morning, the full force of the media's descent on Tanner Drive stoked an ambitious clamor for facts and details.

Ramona Nye, the *Reporter-Telegram* general assignment reporter who with Crimmins had heard the initial call for Jessica over the scanner, watched as a *Boston Globe* reporter tapped on a Tandy 1400 laptop—the latest in mobile computer technology at the time. The screen lay over the keyboard with a small plastic handle jettisoning out of the top. Unfurling the equipment, the *Globe* reporter unclipped the flip-top screen to hammer out a story on a small table in the Sprague backyard.

Nye's eyes bulged. Having worked side by side with heavyweights in her first full-time job out of college, the twenty-three-year-old realized she was working on a national story, the biggest one she would ever cover in her journalism career. She, along with the slew of fellow local media, tried topping their national counterparts.

"We were sending notes in to Jessica's mother to try to get interviews and stuff," said Nye.

One reporter—though it is not known exactly at which point during the rescue—did the heavy lifting, working his way to the front door of the Moores' home where Cissy answered the door.

"No statements," she announced from the doorway.

The reporter persisted with a quick follow-up question, clearly taking advantage of the moment and essentially pleading for a salvageable sound bite to use on the air.

"No statements at all?" he asked empathetically, before quickly going in for more. "Are you related to the baby?"

Without hesitation and striking a tone of confident defiance, Cissy responded, "I'm her mother."

There was a moment of pause, and the reporter needed to think quickly of how to respond to keep her talking and get at least something on the record despite the excitement and shock of talking to the little girl's mother eliciting a rush of excitement. "Oh," he responded. "Could you just tell me what happened—even just that?"

Cissy kept it simple.

"She fell down the well," she replied, appearing resilient but on guard.

The reporter continued an unpracticed and unmannered stammer. "Where.... What—uh, could you just tell me if you think she's going to be all right?"

"She's alive. That's all I know. She's alive," Cissy said, looking the reporter straight in the eye, her hair in strands, a splash of acne on her chin around the corners of her mouth where her round, puffy, and almost puckered lips came down.

The onslaught had begun.

Chip McClure describes a late-night phone call that came in on Wednesday night from the *National Enquirer*, offering money in exchange for their family's story.

"Listen, mister, we're not interested," he said to the reporter on the other end supposedly sitting in the *National Enquirer* offices. "Do you understand? We are not interested! Not in selling you this story or anything else. So, get off this line and don't call back again."

The clamoring for a statement put the police department's spokesperson, Corporal Jim White, under pressure from the media as the desire to hear from the family began to build. At the same time, Chip McClure found himself considering the idea of a statement just as local advertising and public-relations whiz Kenny Karr appeared at the Moore home. Chip recalled not knowing exactly how Karr was allowed inside; nevertheless, Karr was inside the home.

"Hey, buddy, you are going to need me," Karr told Chip. "I'll keep those press hounds from eating you alive."

Chip did not know whether to trust Karr or toss him out on his rear. Desperate and not knowing the potential of the media's force, Chip relented.

Sitting down at the kitchen table, Karr interviewed Chip, asking one question after another and commenting on each response with "Good point." Or, "I don't like that."

CHAPTER 12

By the time Chip finally worked up his courage, it was 10 a.m. on Thursday. He walked to the front yard to deliver his speech to the world; he had practiced the written statement several times as he and Cissy considered the process.

"They were a little concerned about what questions might be posed to them," said Karr. "They were so involved in their situation, they decided they didn't really want to talk to the media, and Chip agreed to speak to the media."

Chip, who had just turned eighteen two months before, was the head of this family, and his youth showed as he stammered.

"With the Lord's help and with your prayers, we know that little girl is gonna make it," Chip said, finally looking up from his piece of paper in his last two words with a grin. He emerged unscathed, having delivered a convincing and evocative briefing.

The national press corps began to gawk at the story, and soon it would not be only a few daily news reporters. Coverage from these reporters started getting attention from the major networks as well as from the burgeoning cable TV station CNN, which had yet to prove itself a news business. As heads began to turn toward Midland and the voice of a little girl coming up from the depths of a dark well, being there to cover that story in its entirety in real time was seemingly impossible. It had never been done before, given the constraints of newsprint space and limited airtime.

CHAPTER 13

JESSICA MCCLURE PROVES OUT CNN'S BET ON CABLE MODEL

The News Business Goes Live

TO TRADITIONAL TV NEWS OUTLETS, THE IDEA OF A twenty-four-hour network dedicated only to news was outlandish. Cable television, though, made niche broadcasting a disruptive force, and only a handful of people in the television business saw the opportunity as cable technology made its way onto the scene in the late 1960s and early 1970s. One entrepreneur among them, however, saw the path to greatness there.

Eyeballs were growing in number in 1979 when Ted Turner and a small team had begun to prove out a model that would support a twenty-four-hour news network. He had not been the only one, however, to envision an around-the-clock news channel. Maurice "Reese" Schonfeld and Turner would put their experiences together to produce their vision of a new kind of news network, with Turner reportedly investing about $20 million into the venture when they launched in 1980. As the first all-news channel to lead into the next realm of satellite-fed cable television, Turner had tested the waters for a while. To make it work, he knew that more would have to come into play than just serving in-demand consumers with news content. In truth, the idea of an all-news network is not the innovation that made Turner's twenty-four-hour news network come to life. Much of the track had already been laid that would help Turner's CNN train get rolling, and it came in the form of the deregulation that made satellite transmission possible.

Before Ted Turner could center his attention on his all-news station, he had cut his teeth on other media. Turner had worked his way up to his position after taking over his father's billboard advertising company following his father's suicide in 1963. Ted Turner was twenty-four at the time. The billboard company was in trouble after having taken on huge debt to acquire a competing company. Still, Turner turned the company around and soon made it the largest billboard advertising company in the Southwest.

Turner was scrappy. Determined. And infectious.

Turner did not want to keep his revenue streams tied up in billboard inventory. He saw more and more of his clients allocating shares of their advertising spending to television. Elbowing his way into the television market was going to be difficult given the limited availability of affiliate stations up for acquisition. Each of the three national networks had an affiliate network in each major market nationwide. Simply establishing a new station in each market to compete with the three networks would be almost impossible, given the challenge of finding enough content to produce and retain an audience. When the FCC allowed UHF operators to broadcast for television audiences, Turner made his move and took over a failing UHF station in Atlanta (then called WJRJ) and another in Charlotte. He changed the name of his company to Turner Communications Group and renamed the station WTCG.

In 1972, new federal regulations made cable television's spread possible when the FCC enacted rules to let UHF operators export their signals to microwave transmitters at cable companies. Up to that point, the FCC's rules had effectively blocked any chance of growth for Turner and other stations to expand their reach through cable operators. Turner's WTCG station had been stuck in the mud with no way to expand the reach of their programming. But now, under the new FCC regulations, his content could be pushed out to a variety of content-hungry cable operators.

But—what content?

The eight years between the UHF deregulation and Turner's eventual launch of CNN would give the cable industry enough time to figure that out. In 1972, a total of 2,841 US cable networks existed, with about six million subscribers, combined. By 1980, on the precipice of CNN's launch, another 4,000 cable outlets had arrived on the scene with about sixteen million subscribers combined.

The intervening eight years was also enough time for Turner and other cable content providers to figure out how they would distribute content to each of the cable operators. That distribution speed bump was eclipsed by an actual rocket launch. On December 12, 1975, a Delta 3000 rocket sparked to life at Space Launch Complex 2W at Cape Canaveral, Florida. Aboard was Satcom 1, the first in a family of satellite communications systems for commercial use, produced by RCA American Communications and RCA Astro Electronics. Hurtling into space with Castor-4 solid-rocket boosters designed for ballistic missile systems, the rocket launched the satellite that would eventually be responsible for a variety of transmissions, including voice, data, and facsimiles to the continental United States, Alaska, and Hawaii. More importantly, it had twice the power of the competing satellite system developed for Western Union, the Westar 1, by the Hughes Aircraft Company. PBS and the other networks used the Westar 1 satellite to transmit their content to affiliate stations. The emerging cable innovators saw the potential in using the Westar 1 satellite. HBO kicked off its satellite operations the year before, but it eventually moved to the Satcom 1 system, which orbited the earth catching incoming radio signals, amplifying them, and beaming them back to earth. Even before he conceived the idea of stretching his legs into the news realm through CNN, Turner had used the space satellite systems to expand WTCG, turning his affiliate into the first "superstation." By 1979, Turner had expanded the station's reach to 4.8 million cable subscribers.

But his superstation's reach was not enough.

When HBO emerged in 1972, Turner saw that networks were going to eventually come to blows with cable TV as a future competitive force. Others were making their way into key niche spaces, too, with channels like The Learning Channel and Nickelodeon. Turner saw that ABC, NBC, and CBS would be looking at how they were going to survive and remain competitive with those cable producers who would be developing content that targeted fractions of their audiences. The result would be a television environment broken down from broad appeal with wide-ranging audiences into narrower niche audiences. There would be a race for those consumer eyeballs to rip revenue out from under the major television networks.

"There are only four things that TV does," Turner explained to Schonfeld at the beginning of their discussions around what cable could do to attract niche audiences. "There's movies, and HBO does that. There are sports, and ESPN does that, unfortunately. Then there is regular series stuff, and the networks do that. All that's left is news."

As networks, trying to compete with cable TV, began focusing on content more appealing to these niche groups, they neglected their news audience. For them, it was a glum piece of the content pie that was unprofitable. Whether news could make money was hardly part of the conversation, especially at CBS in 1987. In June that year, Howard Stringer took over as president of CBS News. CBS News held record low ratings with *CBS Evening News with Dan Rather* and experienced its worst ratings since 1964. Stringer faced hard decisions on where to take the program to dislodge it from its lowly ditch in third place among the nightly news shows.

Options existed. Not on the list of potential solutions: more news.

For the corporate networks, any equation that cut into advertising inventory did not make sense.

"It was traditional at the time for CNN to stay with its live coverage, and that policy started way before I got there in 1990," said Tom Johnson, who would eventually become president of CNN after serving as publisher of the *LA Times*. The Harvard MBA had built a career in news at the highest echelons of journalism—but only after working with his sleeves rolled up in the sleepy community market of Macon, Georgia, for the local newspaper, the *Macon Telegraph*. "Ted Turner decreed that it was more important to do the news than to cut to commercials. That normally would put us against the sales department."

Turner, though, had created a different business model with his news product that was not purely based on consumer demand. His visionary discovery was not just in the realization that consumers desired news content around the clock. It came from being able to do so in a way that network news departments from CBS, ABC, and NBC could not. Turner assembled seventeen big name sponsors to cover advertising, but as part of a cable distribution network, he recruited twenty of those 4,000 cable outlets nationwide to affiliate with CNN. They would each pay CNN 15 cents per subscriber per month. Effectively, Turner once again figured out the importance of diversifying revenue sources and reduced the command and control of advertisers to dictate when, how, and if their ads ran. Turner said CNN's competitive news programming would be enough to lure them. Until then, he said, with a staff of three hundred and 33 percent of the budget of the large network news outlets, he was prepared to lose $2 million per month for eighteen months to make the venture work. He would address the market's latent clamor for news and current events with a two-hour domestic and international news broadcast during prime time. Johnson recognized that while they were betting big that consumers had a bold appetite for news content, it was not a proprietary insight. Their network counterparts knew quite well that audiences would consume more news coverage. Turner, though, had devised a strategic move that would allow CNN the leeway to service that appetite in a way the networks could not. According to Johnson, CNN's emphasis on advertising revenue was not as pronounced as its ABC, NBC, and CBS competitors. For CNN, half of their revenue came from cable operators, and the other half came from advertising. "We were in a far better strategic position than the networks," Johnson said.

While it might be seen as a sure thing in retrospect, the CNN product was still a start-up enterprise. The news team had to get programming in the form of news content. Steve Shusman would be among those hired in the early 1980s to join the team at CNN, and what he saw was a major departure from his experience in large network news operations. Starting in 1964, Shusman had been making the rounds, working his way into several top ten markets like Washington, DC, Philadelphia, and Cleveland. CNN, however, was pulling more youthful and inexperienced staff. "So, it was kind of like we're going to have this experienced management that helps nurture and guide and develop these younger folks," Shusman said.

"I mean, everybody, I mean everybody knew that we were trying to work for a company that needed you to, like, try and get eleven pennies when you squeeze two nickels together in order to make things keep going sometimes," Shusman said, describing the sort of team mentality that develops in a foxhole.

On several occasions late at night before CNN moved into its giant headquarters in downtown Atlanta, Shusman recalls Turner coming down into a break room from his on-site apartment. Their first offices had been cut from a sprawling abandoned country club, and Turner took up residence in a corner of the main building's second floor.

"And he would come down into the atrium sometimes after midnight. You know, like in bare feet, maybe a t-shirt, terrycloth bathrobe, and, you know, he drops a dime into the pastry machine, and he would get those dunking sticks," Shusman said, describing a kind of donut-like sweet snack wrapped in plastic. Turner would then pour a cup of the free CNN coffee he kept on hand for staff around the clock. "He would sit there, and he'd have this crummy cup of coffee in the cardboard cup from the machine. And he opened his package of dunking sticks, and he'd sit there and shoot the bull with the producers who were maybe taking a break in the atrium. Everybody kind of felt that we were all in it together and we were part of this big team that had this goal that we were all going for."

Turner would emphasize the business model in these late-night discussions and focus on the importance of always keeping cameras pointed at the story and keeping them running on air. It was a long game strategy after all, and Turner had spent a lifetime feeling undermined and disrespected with low expectations.

"All my life, people kidded about me," he told Larry King, the late-night talk-show host. "When I sailed, started sailing, you know, I said I was going to be a champion, and people laughed at me then. And people laughed when my father died, and I took over the billboard [business]. I'm used to people laughing. People laughing at me just makes me dig in and work a little harder. It's incentive."

Turner's investment in the business manifested in more ways than one, and his time spent with staff in a show of integrated values for the success of the company coalesced the team. The result was a company where employees saw that leadership was willing to invest in their careers while also working in partnership for a major disruption to the news business. Shusman said, "People who had been there in the early '80s would have probably walked through fire. He ran the kind of a company where it was like a real family kind of atmosphere. It wasn't a corporate nature, it was run like a family business, a successful family business."

Compared to the network roles Shusman had held previously, his contributions at CNN were significant. Previously, despite being at a major network in a large market, there was less autonomy, toxic competition, and little teamwork.

"I felt like they looked at you as basically a space filler. They didn't care who you were—what you were. They didn't care if you worked there the next day. It was that kind of an atmosphere—a real poisonous newsroom atmosphere," Shusman said. "But you have to understand, when CNN was first formed, there were two levels of professionals there. One of them was college kids who were willing to work for experience and, you know, basically peanuts, in order to get the experience. And the other part was the managers who were brought in from top ten market networks. Executive ranks."

Doubt swirled throughout the business. Earl Maple, a CNN video journalist, thought it was clear that CNN was a temporary endeavor. "Nobody thought it was going to work. I thought it was going to be a part-time job until I could get on with ABC Sports," Maple said.

Bernard Shaw, the longtime CNN anchor who became a household name in the early 1990s during the US bombing of Iraq, where he was stationed in Baghdad to cover the story, said that challenge allured him in a way because the stakes were so high.

"To me, the idea of a 24-hour-a-day, all-news television network was the consummate challenge, the ultimate challenge. When I decided to leave ABC News, they said, where are you going?" he explained.

He would outline plans for CNN to the chagrin of his fellow ABC News staff. "I felt like it was the last frontier in network television news, and it was a challenge. People were laughing at me."

Walter Cronkite, the dean of contemporary network news and himself a sort of pioneer who started out in print journalism before making his name as the anchor of *CBS Evening News*, had his doubts. His view on CNN's chances—stark as they were—was shaped less by CNN's news product and more around the investment itself.

"I thought it was a valiant effort that would go nowhere. I couldn't believe that any entrepreneur would put up with the vast expenses of organizing a 24-hour news service and stay with it long enough to make it go," Cronkite said. Doubters and naysayers, though, many of whom had celebrated positions reporting on major issues with authority and almost total control, were starting to feel their own pressure. Sam Donaldson had been chief White House correspondent for ABC News for more than a decade when CNN emerged. In those early days of CNN in the 1980s, Donaldson could still get his news content on the air for a major story, even if CNN initially beat him to the punch.

"When I was at the White House the first time around, I got beat at 2 in the afternoon, I'd still have it at 6:30," he admitted. But as the years ticked on, and CNN grew its reputation and ability to take news straight to air with a stronger journalistic team, it became more difficult to compete thanks to the sheer speed of the twenty-four-hour model.

"If I got beat at 2, by 2:30 Wolf [Blitzer, the CNN anchor] would have it, John King would have it. By 6:30, it would just be old news," Donaldson said.

The combination of being part of a team building a system that was responsible for shifting the news paradigm, paired with youth and vigor and drive, built momentum quickly. Those early CNN settlers describe moving to Atlanta where they knew no one. They developed relationships internally with their news team and spent time together, establishing a social network throughout the company. "So, we became each other's best friends as well as people we worked with," said Jane Maxwell, a CNN deputy managing editor. Others described rafting the Chattahoochee River together on the weekends and gathering for dinners on the town. Scott Leon met his wife at CNN, thanks to the internal social network. "Who else could you find to date working such strange hours? I mean, who else would go out for dinner at 3 in the morning?"

And the underdog mentality, combined with autonomy, drove the team.

"That's what gave all of us this sort of inspiration, was each other," said Ed Turner, a former CNN managing editor. "We were the underdog, we were the fledgling, and people would make fun of us."

As CNN matured, news teams in control rooms—and in the field—were able to move in harmony with authority and ambition to get compelling news covered. "CNN at that time: The story was it. If you had to blow out commercials, you know, you hope you had a really good reason to, but if you had to, you did it, and then you would try and recover later. But there was never, never any thought about making a story secondary. It was always getting it on the air and getting it right," Shusman said.

As Johnson put it, this nature of sticking with the story and neglecting commercial breakaways gave CNN an added foothold over the networks. "For the longest time, it took something dramatic for the networks not to break away for commercials," Johnson said. Bob Furnad, who many said likely took the helm at CNN's main headquarters as CNN's lead producer during the Jessica McClure rescue, said, "That helps solidify our presence to the public when there's a major news event. You know CNN's going to stay with it."

CNN's independence from the tether of advertisers freed reporting teams from worry about whether sticking with a compelling story might draw the ire of programming staff.

Turner's concoction and focus created a powerful cocktail. Bernard Shaw, in retrospect, twenty years after the founding of CNN, told ABC News veteran Sam Donaldson, "On the one hand, Ted Turner was very lucky, Sam, but he was and remains a superb tactician." Over the first decade, cable outlets running CNN would increase their subscribers by 300 percent, going from 17.6 million to 54 million.

At the rescue site, regional newspapers and local television outlets were the primary news agencies on the scene, and those reporters could see that the key attributes of a stellar news story were beginning to emerge.

Dallas Morning News reporter David Marcus said, "I must admit, I'm stunned. It's a very, very touching story, and I think that must be the reason is there's something that grabs you. She's crying. She's talking."

Helping local media along were local police, which allowed local reporters to get closer than reporters from the national outlets. Nye, one of the local *Reporter-Telegram* reporters, as well as KCRS-AM reporter Lisa Barker, used the opportunity to their advantage.

Barker knew about hanging on for a story. She had to be a fighter to survive the news business in West Texas in the 1980s. Barker's victimization came at the hands of the oil bust. Moving to Midland from a weekend anchor position with a television station in Amarillo, Barker knew she needed a smaller market if she were ever to land a weekday anchor position at a six and ten o'clock slot.

Her move to Midland worked, getting her a job at the local NBC affiliate. With the bust, though, the company cleaned house. "Bloody Thursday," they called it. Wanting to stay in the area and tough it out, Barker bided her time, taking a job as a reporter with KCRS-AM, reporting news and toting around a tape recorder and microphone.

Barker's work around Midland and Odessa accompanied her small market television experience, where news production concentrated on bulk. Over the course of the rescue, such concentration resulted in a collection of tapes taken at the scene. She had been there since hearing of the event over the police scanners. One of the first at the scene, Barker's car stayed on the street in front of the home at 3309 Tanner Drive. There was no way she would move it—not that she could. Getting back in would be impossible, and the two-way radio was her link to the KCRS studios. Filing a story became as simple as trotting out of the Sprague backyard to the vehicle a few yards away and firing up the two-way, its crackling transmission pumping out information live from the scene.

Technology had yet to provide reporters the advantages of expensive cellular phones, laptops, or satellite communication. Instead, Barker narrated her stories into the tape recorder. Occasionally, she scribbled the story onto a notepad. For each story filing, Barker hustled to her car and spoke into the handheld two-way radio. Other times, she would just press the play button on the tape recorder, letting the in-studio anchor at KCRS edit it down.

Wednesday night, Barker slept in the Sprague backyard, placing a shirt over a rock she used for a pillow. To rest in between, Barker retreated into the Sprague home where Maxine left the coffee pot going.

Reporters and police sprawled out on the floor, some of them even laying claim to the couch. Other neighbors brought food in for the rescuers, reporters, and police. Dutch and Margie Lunsford, several houses down, dropped off sandwiches. Restaurants regularly swung through with carts of food.

"The police asked me what I thought about the media being in my yard and house," Maxine Sprague told *Reporter-Telegram* scribe Jane Marler Dees. "I told them they've been very nice and quiet."

For the wrinkling quiet lady with a voice as sweet as Jessica's, everybody on the scene was just doing a job—and so was she.

"I'll remember tonight about how nice people are. About how caring people are," she said. The neighborhood's surrounding streets were packed with people: first responders, reporters, neighbors, and their family members. In the rush and clamor, reporters found assistance from others like Sprague. Tony Clark, based in Dallas for CNN, quickly found that their option for covering the story visually was still difficult. Trapped in the Sprague backyard or blocked in amid the myriad of trucks and vehicles in the alley, Clark and his team still needed more and better access to the scene to establish a perspective on the action of the rescue.

"When we got there, I started knocking on doors. And I would say, 'I'm Tony Clark from CNN. We're here to cover the rescue attempt of

Jessica McClure. I need your help. We're trying to shoot over the fence. Do you have a ladder we could use?' You knock on another door and say, 'I hate to ask you this, but can I use your phone?' That was the day before cell phones."

Without Sprague and other neighbors granting use of their phone lines, many reporters would not have been able to deliver the story of Jessica's rescue to the world, whose grip on the little girl tightened with each description of the tot's struggle. It was so real, so vivid—even on the other side of the world. Newsrooms across the United States prepared for constant coverage as the final act of the rescue neared. The producers and reporters standing in news offices and control rooms dropped their guards. Bob Furnad, a CNN producer at the time and eventual CNN executive vice president, saw the power of the visuals and the pathos in the story.

"I could see that this is a really terrific story," Furnad said in a thirty-five-year anniversary program aired on CNN about the organization's evolution. He told the staff: "And it's not going to be resolved quickly. Blow out all the commercials. Stay until it's over."

Their grip on the rescue resulted from their own investment—or perhaps it was the opposite—as reporters found themselves hoping for a positive outcome. The thought of having to cover a negative outcome would have been worse—perhaps several times worse—after the buildup of hope among so many who believed the impossible could indeed be achieved.

It was a major shift in their perspective as some of the reporters from Dallas and other cities started to arrive on an assignment they did not totally assess in advance of their arrival.

"Initially we got the call and said we were going, I thought: Well, you know, a girl in a well. You know, she'll probably be out by the time we get there. I just didn't realize the significance of it," said Gary Utlee, a cameraman with WFAA-TV in Dallas. What a boots-on-the-ground perspective revealed, however, was the enormity of the rescue operation in an otherwise nondescript neighborhood. Greg LaMotte, a CNN correspondent, felt the story build with tension as the rescue dragged on. "And then as a few hours turned into 10 and 11 hours turned into 12 hours and 14 hours, you began to think that, or at least I did, that she wasn't going to make it."

As Thursday toiled on with the constant drone of the Green Dragon and drillers emerging from the rescue shaft in much worse condition than in the early morning—covered in much more dust and hacking on the particles while also trying to flush their eyes with water—the reporting stalled out. With only so many stories produced at the scene, reporters branched out, combing the neighborhood as their desperation for an original angle became obvious.

One *Dallas Morning News* reporter looked on as a reporter from a newspaper in Hawaii questioned one of Dutch's grandchildren playing on their living room floor just a few feet from where rescuers took breaks. The Hawaii newspaper reporter started working on an angle, asking the grandson about his experience with the McClure family. A few weeks earlier, the child had sold candy to Cissy for a school fundraiser, while a handful of children scampered around at her feet.

Others took to more solid story leads, concentrating finally on why the well in this backyard remained open, uncapped, and just waiting for an accident like this one.

Austin American-Statesman reporter Debbie Graves dug into the regulatory aspects of the open water well, hunting down Texas Department of Human Services official Susan Hollan, who quickly concluded the well should have been capped. More importantly, Hollan told Graves, the daycare was not legal in the first place.

"We had never been out there," Hollan told Graves. "We didn't know about it until the child fell down the well."

Considering that home childcare facilities face higher scrutiny than many other independent businesses, registration was key—the well would have been found and capped for the danger it posed had Texas's DHHS (Department of Health and Human Services) known about it.

CHAPTER 13

A regional daycare licensing director told Graves that for every two to three unregulated facilities, there is just one complying with regulation guidelines. Graves's reporting revealed that the regional licensing director had only two investigators and six licensing representatives to cover twenty-three counties in West Texas—45,000 square miles, an area roughly the size of the state of Mississippi.

Being a journalist in West Texas means developing angles from national stories to generate a localized account of the same issue. Reporters call it "feeding the beast": the beast being the newspaper's massive layout, and the feeding being the slew of stories reporters churn out—at least two and sometimes four to six each day to ensure adequate local coverage. Learning how to "localize" a story in Midland yields a higher probability of success. This is the practice of taking trending national and international issues or stories from the wires and interviewing local sources on the same issue for local context and perspective.

But suddenly the tables had turned.

Newspapers and TV stations from around the world developed localized angles, along with historical reminders of similar stories from long before, based on what was happening in West Texas. Perilous stories with comparisons to Jessica McClure's situation suddenly emerged as part of the news coverage worldwide.

In the small Italian town of Vermicino, Italy—just outside of Rome—locals recalled their own struggle to save Alfredo Rampi. In 1981, Rampi had fallen 196 feet down a well shaft. For four days, rescuers worked to save Rampi while cameras rolled. But Rampi died when sand filled his lungs.

"Jessica is like little Alfredo there, underneath, a little voice singing," wrote Rome's daily *Il Messaggero*.

A more similar rescue to Jessica's was that of Kathy Fiscus in 1949. Fiscus had been running in a field in San Marino, California, on Friday, April 8. Falling 149 feet down a 14-inch well, the three-year-old clung to life.

News media swarmed the scene, and reporter Stan Chambers of KTLA, Los Angeles, stayed there for thirty hours until they brought the little girl up—dead.

"They brought the girl's body up. It was terrible," Chambers recalled of the struggle.

The weed-covered well had claimed the girl's life in spite of the hope and prayers of the 15,000 who gathered at the scene watching rescuers. Others watched live on the pioneering KTLA television station—the first to explore the magic of live television news and its impact on drama made from nonfiction.

KTLA knew of the power of live news. They tested its strength, becoming a station to take many "firsts" on the West Coast, with the claim of being the world's first on-the-spot news coverage on February 20, 1947, at the Pico Boulevard electroplating plant explosion.

The legitimacy of news generation took reporting to a higher level. The power of television surged with the ability to generate images—live images as they unfolded. On Tanner Drive, most local stations broke into live coverage as soon as word came that the drillers finally pierced the well shaft that held Jessica.

At 4:30 a.m. on Friday, anchors told the world, rescuers drilled just inches away from where Jessica dangled waiting to be saved.

CHAPTER 14

COPS KNOCK ON THE COP'S DOOR

THE TEXAS RANGERS AND MIDLAND COUNTY SHERIFF'S deputies, together with police from Odessa, rapped on Glasscock's front door in a community on Midland's outskirt town of Greenwood. His buddies, Texas Rangers Jess Malone and Don Williams, wanted to talk with him. It was Christmas Eve 2003. Glasscock was off duty that day and eating lunch at Chili's, taking a break from Christmas shopping with his youngest daughter and ex-wife, Lynne. Their relationship's improvement over the previous months had made life bearable for Glasscock and Lynne.

The 2001 fallout over the revelations about the Paula Bynum incident had resulted in their divorce. A month after the shooting Lynne had moved out and into a home a block away. Two years on now, things between Glasscock and his second ex-wife were finally getting better, leading to an occasional romantic liaison at his home. The relationship, however, would remain that way only for a couple more weeks.

Michael, Glasscock's oldest son, called his father's cellular phone, sending a pinging ring around the lunch table at Chili's.

"The sheriff's office called. They want to talk to you," Michael told him.

"What do they want?" Glasscock asked.

Michael answered, "I don't know, but you better come out here."

Before Glasscock could drive the ten miles home, police vehicles had swarmed his house: Odessa Police, Midland County Sheriff's deputies, and Texas Rangers, who traditionally called out when investigating other law enforcement entities. All of them could be counted among Glasscock's acquaintances and friends.

CHAPTER 14

In a back garage, Glasscock stored a collection of munitions that were supposed to be on their way to storage at a designated facility owned by the police department. But he had kept 400 pounds of high explosives, 1,700 blasting caps, an artillery shell, two pipe bombs, and four homemade hand grenades on the property instead of sending them on to the storage facility. Still, when his son called on Christmas Eve, Glasscock did not know what to make of the call.

"You're under arrest," Glasscock heard Malone announce as he walked up his front lawn, though no one grabbed him for handcuffs or physical detainment. Glasscock glanced up to see a news crew on the scene. A KWES–Channel 9 camera was nearby and narrowing its focus on him and the confusion in his driveway with the swarm of investigators. Glasscock's sons waited on the driveway, too, as officers kept them clear of the house.

Lynne stayed behind at Chili's to finish lunch and Christmas shopping. She thought nothing of the call summoning her husband. Her cell phone beeped for her attention; it was Michael on the line with the news of Glasscock's arrest.

"For what?" Lynne asked, her breath leaving her for a moment as shock set in.

According to Odessa Police investigators, Glasscock had forced a woman to have sex with him just four days earlier. And, they said, he had supplied his victim with a drug. Local news outlets would not report the victim's name—and even later, in July 2005, when Glasscock pleaded guilty to charges of sexual assault, taking a twenty-year plea bargain with Odessa's district attorney, the woman's name still remained elusive to the public. Investigators would allow only a pseudonym to identify her. According to police, "Sandy Lou Smith" was a fifty-year-old Odessa woman. She told police that Andy had taken her out for a few drinks the night of December 20 before taking her home and giving her medication "without her knowledge, rendering her unconscious." Even Lynne, who knew of Glasscock's infidelity, did not believe the charge when she first heard about it. She knew the accuser well—even though law enforcement used a pseudonym to protect her identity from the broader public.

168

Looking back over her marriage to Andy, Lynne found the accusation unbelievable. Sandy Lou Smith's other accusations made Lynne skeptical of the validity of the rape accusation. But Smith had a kicker: Glasscock had videotaped the encounter.

By the time Lynne arrived at Glasscock's home from Chili's on Christmas Eve 2003, at what was now labeled a crime scene, her ex-husband was being carted away by police—no one was allowed in the home.

"I'm gonna go home and drink," Lynne thought to herself, overwhelmed by exhaustion. Flustered and shocked—half in and half out of her mind—she backed her vehicle out of the driveway, dinging Ranger Williams's vehicle.

Glasscock later claimed the sexual assault was actually a consensual encounter. The drug he gave the woman, he said, was Ambien, which they each took after splitting the pill in half. When Glasscock made bond later that day, he said nothing. His skin was pale. His confidence waned, and instead of vowing to fight the charge, he recalled what he had seen in his home when officers arrested him: investigators collecting his computer and videotapes.

"What's wrong?" Lynne asked, attempting to coax Glasscock from his rut. Afraid Glasscock may be planning suicide, Lynne had moved him into her daughter's bedroom for the remainder of the Christmas holiday.

"I was afraid he was going to blow his brains out," Lynne said.

He had thought about it, but not long enough to spur any action. All he could do was answer questions about his new depression and claim stress had overwhelmed him.

"I'm ready to just chuck it all in," he spouted. "I don't want to do this anymore."

Lynne lashed out.

"No. You're not leaving the two kids without a father," she told him forcefully. "You're not doing this at Christmas."

Glasscock moaned, forcing his face into his hands when he was not staring at the ground in disgust with himself.

"It's just gonna get worse," he said.

169

CHAPTER 14

Glasscock's collection of pornography was well known around the police department. Officers even stopped by to borrow material, and Glasscock handed it out like a willing librarian. Before he and Lynne split up in 2001, the two argued over the porn he downloaded from the Internet. Although the material she came across included "legal" forms of pornography, she made her displeasure about it being on the family computer known frequently.

"If I found it on there, it got deleted quick," she said.

Lynne thought about her past arguments with Glasscock more than once between the time of his arrest and the New Year. Her support for the father of her children held fast as she scoured their past for memories of Sandy Lou Smith rather than dwelling on the computer.

When school resumed from the Christmas break, Lynne's vision of the arrest was more gray than black and white. Her daughter approached her with confusion after some of her friends were questioned by state authorities at school.

"Mom, why is Child Protective Services coming and talking to all my friends?" Lynne's daughter, then twelve, asked in early 2004.

Good question, Lynne thought.

"Why is that, Andy?" Lynne passed along the inquiry.

Glasscock sighed on the other end of the phone line. "Well, it may have something to do with the tapes," he began.

Lynne's teeth began to grind, her jaw clenching. "What . . . tapes?"

Andy paused, sighing again, his voice sounding squeaky and nasal. "The tapes I made," he answered.

"Of?"

"The girls in the shower," Glasscock explained.

Lynne paused for a moment, squeezing her eyes tight. "What . . . girls?"

Getting spy cameras set up is a simple process. Being a police officer makes it even easier. But when Malone and Williams combed through Glasscock's home, they found one camera placed in a clock radio in the bathroom. They found another in Glasscock's bedroom.

"Is my daughter in any of this?" Lynne asked investigators during her interviews with police, Rangers, and agents with the Federal Bureau of Investigation.

COPS KNOCK ON THE COP'S DOOR

No, but her friends were, Glasscock admitted. This was confirmed for Lynne by investigators. The next day, Midland attorney Tom Morgan secured a protective order to keep Glasscock away from his children.

As Lynne spoke with an FBI agent, the sudden realization of the implications of her recently improved relationship with Glasscock and their resumed romantic relationship came into view.

Was she also in some of those videos?

Picturing herself in the videotapes, she realized they were now being examined closely by the very agents and law enforcement officials she had known most of her adult life. The embarrassment overwhelmed her, and all she could picture was those investigators watching her in the most vulnerable of intimate positions with her ex-husband.

CHAPTER 15

THE WALLS COME TUMBLING DOWN

THE BREAKTHROUGH INTO JESSICA'S WELL CASING CAME as miner Dave Lilly powered a bit into the rock. Worried that one of the other drillers might drill right into the girl, Lilly said he took control when rescuers knew with certainty that merging with the well shaft was imminent.

The impending breakthrough caused a flurry of excitement among the drillers at 4:30 a.m. on Friday, October 16. As they neared the 48-hour mark, tension was high. Nerves were frazzled, and the volunteers who had poured into the backyard had little perspective to manage the anxiety that can come with a real rescue situation.

"The first one to reach her, just pull her out," O'Donnell heard one driller say. He shook his head, worried one of the untrained volunteers might hurt Jessica further. Any jostling could severely exacerbate any injuries she had sustained in the initial fall. Some of the drillers might understand that risk—but not all of them would have enough of the basic first aid experience to know that any jerking or pulling could have a terrible net effect.

Although the breakthrough meant they were within reach of the girl, a coning effect in the tunneling left a tight space for paramedics to fit through in those final feet of the rescue shaft. A stand Lilly had configured late on Friday night for the drillers to rest their drills on helped with fatigue, but narrowing focus naturally encouraged the diggers to plunge ahead, aiming for the center of the tunnel and a straight path to the well.

The result was a tunnel too tight to let anyone reach Jessica even though the well had been penetrated.

CHAPTER 15

As late as 10 p.m. on Thursday, O'Donnell and Forbes stood by, waiting to go into the hole. O'Donnell hunched forward as Forbes stood with crossed arms and occasionally swatted at the dirt with his foot. It was now 11 a.m. on Friday, and the two were still standing around—wearing the same clothes they had the night before as they waited.

Not knowing how Jessica hung in the shaft or whether the breakthrough might jar something loose—perhaps clearing a path for her to drop farther down the well—Lilly hammered a long iron bar horizontally through the well casing, just beneath the point where the rescue tunnel and well casing intersected. That would hold until he inserted two inflatable bags into the well—one above and one below the level of the rescue tunnel. These bags, once inflated, would fill the space and effectively seal off any route beyond their round mass.

At 11:02 a.m., Lilly dropped into the hole. He disappeared without a sound. KMID-TV had begun an all-out live feed without breaks just an hour earlier. Rodney Wunsch stood by with the microphone waiting for news to break, occasionally leaving the camera with reporter Gayle Hill to fill in while he gathered details. A glare on their TV monitor in the Sprague backyard limited how much they could actually see.

Lilly emerged at 11:10, sending another driller down, who stayed below, chipping away until 11:28 a.m. When he came up, another replaced him. It seemed to be going on forever.

Local KMID anchors straggling into their studios were just getting caught up with Wunsch's coverage when they went on the air, haphazardly going back over information Wunsch reported moments earlier.

Can you give us a timeline? they would ask again and again.

Wunsch squinted back into the lens. "You can't even imagine how many times that's been asked and how many times we get the same answer. They don't really want to tell us anymore, because every time we give a time, it's something to shoot for and something to be hopeful about. Time and time again, it hasn't come to pass. I believe the last time we asked Corporal White how long, he turned around and walked off." White had also grown weary of the constant and redundant media inquiries.

"Most of them acted very professionally and stayed within the confines of the media compound," he said, but with each new wave of reporters that had arrived on site with fresh eyes and editors hounding them for updates, White found himself in a constant cycle of repeated statements and facts. Having left the site only once for a break, White did not get a full picture of just how big the rescue story had become. "I knew it was escalating daily, but at the time, I didn't realize how big it got," he said.

One thing he did not realize was that as the spokesperson, preparing for the worst outcome might be worth considering. He had yet to develop a statement to the media should Jessica not make it out alive. "I never even thought about it," he said.

At 11:41 a.m., Glasscock rose to his feet, raising his body with his hands first at his knees then at his waist before rising and stretching his back. He squinted into the light that finally hit him in the face. His dark pants were gray with dust, his face was covered in the same. Lilly sat nearby, cross-legged and waiting for a driller to come up with news of his progress.

Reporters looked on, the stillness leaving them lapping up any morsel they could get their hands on as anchors did the same, trying to narrate to the beat offered by cameramen operating for better angles. The anchors filled the gaps like starstruck kids trying to get a word in edgewise.

"There we go. We know that guy," said one of KMID's anchors like an overzealous autograph seeker at a KISS concert when Czech walked into a frame. "That's Chief Richard Czech." Her voice wanes slightly. "Chief of Police, Midland Police Department."

The news crews perked up as Glasscock and a number of others around the well extracted the heating system, threading the long PVC pipe out from the well. At 12:06 p.m., a dusty driller emerged and stumbled away, hurriedly trying to detach himself from the towline. Another driller went back down the shaft, though, a minute later.

The movement around the well along with the fifty-hour buildup prompted photographers to pick positions. Though many had positioned themselves on the ladders in the Sprague backyard for a number of hours, *Odessa American* photographer Scott Shaw looked for another angle for his shot, hoisting himself into the sky with the help of Bentley's cherry-picker bucket. Though the angle was clearly much too high, he realized it was too late to get picky when at 12:11 p.m. Forbes and O'Donnell began moving to the well with a small backboard in hand—cut to fit the size of Jessica's body.

O'Donnell stepped up to the rescue line. As he did, Forbes patted him on the back. A driller, emerging a minute later, unclipped from the line, immediately making way for O'Donnell. This was the moment the world had been waiting for. O'Donnell and Forbes appeared somewhat cleaner and were not caked in dust. The two were clearly paramedics, though safety lines and harnesses covered their bodies.

"Judging from the patch that was on his arm that said EMT on it, it looked like that was an emergency medical technician paramedic patch that was on it," Wunsch stammered, his reporting under the influence of two and a half straight days of reporting with only minimal rest.

O'Donnell sat on the edge of the rescue shaft, dropping his legs in the hole. Wearing a gray EMT uniform and a green and white mesh cap, O'Donnell relaxed with as much anonymity as any other firefighter on shift that day. Reporters like Wunsch had no clue who he was. Forbes was the same. His EMT uniform, though, was covered by a shiny blue jacket that cut close at his waist, squeezed tight with an elastic band.

News cameras focused on Cissy, standing near the hole, looking on as she responded to them occasionally—clearly talkative and waiting anxiously to see her child.

Five minutes passed while O'Donnell and Forbes waited for Lilly to inspect the opening leading to Jessica. When Lilly emerged, the two abruptly went down the rescue shaft, disappearing as interchangeably as any of the other drillers.

Tony Clark, broadcasting live for CNN from the scene, reported what he could in real time. "We're told that two paramedics are now down in the shaft. And if that is the case, it may be just a matter of time now before they are able to pull young Jessica McClure, the 18-month-old, out of this hole that she's been in for more than fifty-two hours now," Clark reported.

On his way down the shaft, O'Donnell felt the pressure building, making his guts feel as tight and hard as the lines he held onto lowering him into the shaft.

"I don't think anyone can prepare themselves for something like this," O'Donnell later said. Although the firefighter-paramedic's usual experience was with car wrecks and fires, he now had the life of a little girl in his hands.

"We've been handed some information here that the last paramedic in the hole is Steve Forbes," KMID anchor Mike Barker attempted to stream in to Wunsch.

Wunsch replied, "No, I didn't recognize the name or the paramedic. Otherwise, I'd be able to describe them to you."

This felt like the moment for rescuers to finally emerge with the baby in their arms. Cameras were fixed to the site as reporters filled the air with any information they could.

"CNN wanted to stay with it and not break away from it and not say, we'll be back to that story. They wanted to stay with this story. And there was always this anticipation that the little girl could be pulled up out of the hole at any minute," said Greg LaMotte, the CNN correspondent. But planning had already been under way for how they would cover the girl's eventual emergence live on the air. Ted Turner's mantra of staying with the story, no matter what, rang in the team's ears. Stay with the story. But now, more importantly, they had decided to let the camera tell the story, directing the team to do little to narrate as she emerged.

Chip and Cissy McClure stood outside by the well, watching rescuers from behind a line. At 12:34, Cissy suddenly turned and stomped away. The rescue lines started to move, and Forbes emerged still with the backboard—though it was empty. The latest reach into the well had failed. No Jessica.

Then at 12:38 p.m., O'Donnell emerged from the rescue shaft, his hat crooked, his brow covered with a thin layer of sweat. Roberts, Czech, and Lilly quickly surrounded him, listening as he described his position in the hole. Though Lilly and the other drillers had worked to widen it, the hole's restricted tightness limited access to the baby. The access tunnel measured only around twenty inches wide, and O'Donnell slithered his way into the tunnel, scooting along on the hard rock. The tunnel closed in on him and his eyes adjusted, but the hole really was very tight.

When O'Donnell's head hit the end of the tunnel, where it made contact with the well casing, the opening narrowed to just twelve inches. He could barely slide his left hand along his chest, squeezing it hard against the top of the tunnel to reach into the pipe. At last, he was able to feel Jessica's leg, as soft and magical as any baby—as his baby. With Forbes a few feet away, standing by at the bottom of the rescue shaft to assist, O'Donnell explained the only viewable portion of her body was her left leg. It dangled low at the level of his face. Her diaper squashed and squished, now full of two and a half days' worth of excrement. There was no budging her. The stench was overwhelming. She was as wedged in the hole as she was before the rescue tunnel had pierced the well. Touching her, though, brought a level of crazed anger and disappointment. He felt the failure wash over him.

"I'll be back, Jessica. I swear I'll be back," he said, scooting back toward Forbes.

After describing what remained for the drillers to do and telling the chiefs and Lilly about what he had found, O'Donnell retreated from the backyard; he took a seat on the edge of a curb and wept. Seeing him crying like a failing father, several of the doctors on the scene began to question whether O'Donnell should be allowed back down into the well again. One of the pediatricians watched as O'Donnell broke down and wondered whether he should go back after experiencing such a significant emotional mood shift.

Those who knew O'Donnell best say this was the moment that changed his life. He would never again be the same person.

O'Donnell called his wife, Robbie, at her workplace.

"She's right there. I can't get her," he said, shaken and crying. "Everybody's depending on me. And I can't get her."

He was moved in a profound way. Expectations and fear had grown. The emotion in O'Donnell was raw, and it now resonated around the world as CNN transitioned to wall-to-wall coverage at the rescue scene.

"As the hours went on, you thought the chances of her surviving were less and less," CNN's Tony Clark said years later in a recollection of his reporting from the rescue scene. The pitch of the excitement was clear. CNN knew viewers were tuning in. As other outlets broke away to cover the regular programming needed to garner the sponsor support tied to scheduled shows, CNN focused in.

For another hour and a half, rescuers worked as they had been with the drills and hammers, chipping away at the rock to widen the tunnel. Meanwhile, O'Donnell and Forbes stood by, expecting to take another run into the rescue shaft and tunnel.

At 2:15 p.m., the hydro drill appeared on the scene as a City of Midland water tanker made its way into the backyard already full of rescuers and workers.

From what Bill Jones knew, the hydro drill could shoot away rock. Corporal White explained it would be used "to try to erode the rock surface away."

For another five hours, the hydro drill sprayed water with the force of 30,000 pounds per square inch onto the rock, washing the pieces of rock bits away, speck by speck.

At a Holiday Inn just blocks away, Grace Ramirez, a housekeeper at the hotel, turned on the TV in each room she prepared to stay abreast of the coverage and rescue effort.

"Ever since this happened, at home and at work I watch," she told the *Dallas Times Herald* staff writer Bryan Woolley. Ramirez said that earlier that morning, she had turned off TV news coverage of the rescue, only to turn it back on at five a.m. after tossing and turning. And millions of people around the world waited, just like Grace.

People all over the state felt a connection to Midland's rescuers. Alice Grant, a district secretary for Pride Oil Well Service's Kilgore office, watched Glasscock on television as interview requests continued to pull him away from the well after the first rescue attempt. She remembered

Glasscock from their days together at Big Lake High School. Grant felt pride swell inside her as she watched Glasscock talk about the effort.

At the same time, Glasscock's first wife and their children, Michael and Jennifer, watched him on television news from their new home in Bryan, Texas—more than six and a half hours away.

In Atlanta, Earl Maple, the director for that evening's CNN coverage arrived at the CNN headquarters to news that Baby Jessica's rescue was imminent.

"I was doing the 8 o'clock and the 10 o'clock broadcasts, and when I got in, they said we think they're going to get this little girl out of this hole tonight," Maple recalled. When Maple reached Tony Clark, still reporting at the scene for CNN, Maple simply instructed Clark to "Let the pictures tell the story." At the scene itself, rescuers were less positive, having tried and failed so many times before some unforeseen barrier would emerge and make the final step of the rescue impossible. Contingency plans continued being considered.

Around 6:45 p.m. Ronald Short, the man with no collarbones, was considered as a replacement for O'Donnell. Curious as to whether his collapsible body could reach the girl, Czech and Roberts were not totally sold on the idea of a non–first responder taking a position in the rescue tunnel. O'Donnell, who waited in the wings, was still under close observation as doctors considered two major factors that could potentially affect the success of the rescue: Could O'Donnell handle more pressure? Would he have to break Jessica's bones to get her out?

The timetable began to collapse all around them. If O'Donnell again descended into the shaft and could not accomplish the rescue—at any cost—Dr. Chip Klunick would be the one to go down.

At 6:30 p.m., Lilly called for the paramedics after inspecting the opening leading to Jessica. Hopes again began to heighten around the well.

Like they had before, O'Donnell and Forbes prepared for a descent into the rescue shaft. O'Donnell met Lilly at the bottom, his heart racing, his stomach tied in tight knots, and his mind unable to concentrate as the tunnel closed in on him. His arms itched and burned, with adrenaline pumping through his veins.

Lilly helped O'Donnell onto his back, aiding him for a moment to help him slither into the tunnel. Already, O'Donnell's confidence increased. He felt better about the width of the shaft, crawling along with slightly more room—though it remained tight. He said later he was ready to "fight for this child's life."

He would have to—depending on how she was wedged in the well. Feeling Forbes at his feet, he reached up toward Jessica, her single leg still dangling.

"I need something long to adjust her or see how she's stuck," O'Donnell told Forbes.

Down came a photographer's extendable monopod—a device with a telescoping shaft that could be used to hold a heavy camera. It worked like a tripod but had only one leg. Using the rubber tip on the end, O'Donnell worked it around her body, trying to figure where she was stuck. It went all the way around her without catching.

To counteract the sticky residue on the well walls that he could only compare to tar, O'Donnell began rubbing the walls of the well casing in K-Y jelly. Sweating and exhausted, he stretched his arms to slather the well casing with the slimy goo.

Giving her foot a tug, he pulled off her shoe. O'Donnell heard her talk back.

"No," she said as though she were going to kick him away.

The stress of her plight had taken its toll on her, and there was no way for him to know exactly what she thought about being at the bottom of a well for two days with clawing monsters roaring all around her and now feeling a clawing grip at her foot. Her hands, resting near her cheeks, had rubbed along the sides of her head, the sticky tar from the well getting caught up in her fingers. Wads of her hair had gotten stuck in the tar and had been pulled out at the root when she moved her head.

"Come on, Juicy. I got ya,'" O'Donnell called gently, knowing the nickname would calm her and possibly make her believe he was family.

Tugging at the foot, though, seemed to irritate her, so he tried tugging on her pants. They unbuttoned and slid from around her slimming waist. Tugging on the pants more, they caught at her hips, and she slipped slightly into the K-Y jelly.

"I knew I was going to get her out," O'Donnell said later.

Like stepping on a plum, O'Donnell said, she slipped out more.

As she did, Forbes reported bits of their progress, though on the ground, rescuers did not know what to expect. Glasscock did not know what to think as his microphone went dead. The place where the tiny voice of frustration had leaked out of a tired baby suddenly fell silent.

Glasscock thought she might be dead.

Cissy had been told to wait in the ambulance where paramedic Toby Partridge sat at the wheel. Chip waited in a police car with sister-in-law Jamie. Their excitement had spiked before only to result in disappointment, but this time seemed for sure as Chief Roberts told his men to line the alley to create a corridor of protection and carve out a path to the ambulance.

There was plenty of time to organize a safe route. Forbes took Jessica in his arms. He had to cut her hands from her hair where the tar had virtually glued her hands to the sides of her head. Forbes wrapped her head in gauze and the two began to gag.

"What in the world? Are you guys okay down there?" Chief Roberts called over the radio.

"Not really, chief," one of them answered. "We're about sick down here."

Chief Roberts, perplexed, asked, "Sick of what?"

Jessica had filled a diaper for fifty-eight hours, leaving a rank stench of excrement at the bottom of the sultry shaft.

Dan Rather interrupted his normal broadcast on CBS. Other studios did the same. CBS News Vice President David Buksbaum stood in his newsroom as a full staff watched, waiting to see some sign from their cameras that Jessica was coming out alive. NBC News's Tim Russert also stood by in a full newsroom as normally busy producers stopped to watch and wait.

Across the state in Houston, police officers stopped by the pressroom at Houston Police Department headquarters to get updates from *Houston Post* reporters.

"We've gone through every layer of rock with those drillers," said Lieutenant Chuck Lofland, described by *Post* reporter Robert Stanton as a hardnosed homicide investigator.

At the same time, customers and employees of Highland Electronics in Houston stared into the glass line of televisions along a wall, waiting to see whether the girl would make it out alive.

The rigging seemed to take forever to move.

Reporters on ladders in the Sprague backyard watched as photographers waited with their fingers on the trigger. *Odessa American* photographer Scott Shaw abandoned the Dimension Cable cherry picker, opting instead for a spot next to Nye and Barker on the flatbed of a truck.

Weighing his shot, Shaw considered his best bet for a good angle. The area was packed with rescuers around the well, though many moved from the site only to take a place lining the alley leading to the ambulance. If she came out, capturing her and the reactions of the rescuers immediately would be crucial. Shaw and the photographers around him would have a split second to take their best shot. Wanting to get the faces of the rescuers—as well as Jessica—Shaw considered how tight he wanted the shot.

His mind sputtered, having slept for only four of the last fifty-plus hours. And most of that shut-eye came at the rescue scene. Considering that the press room would be past the deadline for a color photo, he loaded his camera with black-and-white film, setting up a flash though enough lighting engulfed the scene to provide a sufficient level of light. The rescuers, he thought, were a big part of the story. It should be wider, he determined, attaching a 180-millimeter lens while other photographers snapped on 300-millimeter lenses to get tighter shots.

The cables began to move as the Green Dragon cranked to life with a hum filling the air as rescuers seemed to hold their breath, watching the cables bounce side to side and slide upward.

Rick Wood, the future general manager of KWES-TV in Midland, had given up on trying to keep his attention solely on the rescue. He felt like clearing his mind of the madness, so he was watching a Midland High football game, listening to coverage of the rescue on the radio.

John Foster, KMID's general manager, who had committed so much to their station's live effort, watched and waited, standing near the ambulance.

David Eyre, a screenwriter for a number of made-for-TV flicks, had checked into a hotel room in Seattle and was watching the rescue. He and a friend looked on, waiting to see the outcome as the lines moved with one of the workers' gloves guiding the wire up.

Earl Maple, the CNN news director stationed in Atlanta, directed Tony Clark's reporting at the scene to allow the moment to emerge naturally, without Clark's voiceover. "And then the instruction was simply to be quiet. Let the pictures and the emotion tell the story," Clark said.

Despite the moment's certainty, worry remained. Since the early afternoon on Wednesday, expectations for a short rescue fluctuated. Ted Koppel, anchoring coverage for ABC News, described the moment well.

"It looks as though, and we have been hearing, such predictions now for so long that I don't want to raise too many hopes . . . but it looks as though Jessica McClure is about to see an end to her ordeal," Koppel said before correspondent Mike Von Fremd took over reporting live from the scene. For more than ten minutes, Von Fremd filled time as cameras caught images of Glasscock, a look of exhaustion on his face. Roberts meanwhile paced around the hole bending over repeatedly to look down the rescue shaft. Workers in hard hats watched from perches on the Green Dragon.

"I remember following the story the day after, and we were told continually the next few hours, the next few hours. Well, as you know, the next few hours have gone on and on, because when they drilled, they found the hardest darn rock I guess you could ever find. Hearing the child cry while you're drilling for two days—more than two days—is absolute [sic] just agony," Von Fremd explained.

The cables tethered to the rescuers wobbled slightly. Taut, they moved from side to side as a gloved hand reached to steady them and guide their movement. "You can see the cable coming up. Everyone, everyone's eyes are looking down," Tony Clark reported for CNN's live coverage.

In a quick burst, a cheer erupted from the crowd still surrounding the rescue shaft. A split second later, Forbes suddenly emerged with Jessica over the rescue shaft's lip. With a bright red trucker hat, Forbes seemingly popped from the shaft suddenly before finding his feet under him with Jessica bandaged and tight in his grasp. O'Donnell, though, was still down below looking up at Forbes's boots before starting to crawl back into the tunnel to gather up leftover gauze and medical supplies. But when he unexpectedly heard the commotion as the crowd surrounding the well burst into cheers, he realized he had not expected anything other than a rapid rush of the little girl away from the backyard toward the waiting ambulance. He had yet to process any thoughts about how the anticipation around the world had grown as much as it had.

Forbes had spent several minutes down below with O'Donnell, trying to tighten Jessica to the backboard that had been cut and downsized to fit through the rescue shaft. The anticipation built through his painstaking effort to secure her to the board, wrapped in swathing white bandages and gauze. Meanwhile, O'Donnell waited in the diagonal rescue tunnel. For several minutes, they had conducted their work as EMS paramedics, assessing the health of the girl, bandaging her, and preparing her for physicians to take over. They had yet to bask in the glow of the achievement, although the little girl was now somewhat safely in their arms.

As O'Donnell heard the cheers, he pulled himself clear of the rescue tunnel and looked up to see Forbes being untethered from the carabiner that attached him firmly into the wire spooled by the rathole rig. Hearing the cheers from the crowd, he yelled in excitement too.

"But I don't think anyone heard me," he said.

In the seconds between Forbes stepping onto solid ground and Jessica being released from her own rescue line, Shaw clicked off sixteen frames with his 180-millimeter lens, waiting a second after each shot for his flash to recharge. In one frame, one rescuer is looking at Jessica's bright round eyes while others are looking away. In the next, a different rescuer is peering at her. In one photo, a shot he thought would be nothing and not particularly the best, he captured a single instant of several men looking right at their goal—her eyes big as hubcaps and staring straight into the sky.

CHAPTER 15

At 7:58 p.m., more than fifty-eight hours after Jessica fell down the well, she was wrapped the arms of Bill Queen, a firefighter-paramedic who hustled her off to the waiting ambulance.

Shaw clicked as Queen took her, thinking the shot lacked focus.

There's no way I got the shot, he thought.

But that shot made the *Odessa American* front page the next morning, taking up most of the page above the fold, Jessica's eyes dark, large, and round peering up into the night from Queen's arms. Tufts of her hair, torn by her own hands from her scalp in frustration, wrapped in a glued mess around her fingers. The tar substance O'Donnell had struggled against when he tugged on Jessica was smeared all around her face and stuck her fingers together in a matted mass, mixed with her own hair.

Shaw's second shot, the one where rescuers all at once looked at Jessica as though Shaw asked for a pose and snapped the shot in an instant, ran on the front of the metro section.

Rick Wood, still at the high school football game at Midland Memorial Stadium, watched as an announcer began the words, "Jessica McClure has been rescued." A hush fell on the audience, and in the distance a line of car horns began a wave across the city.

For a second, there was a way to hush the power and strength of Friday Night Football in West Texas. Foster, Wood's boss, rushed down the alley chasing the ambulance as it pulled out onto the street. He fell to his knees and began to cry.

David Eyre, the man who would eventually be chosen to pen the screenplay for the movie about the rescue, slapped his buddy on the back as they watched the rescue on a hotel TV.

"Yeah! Yeah!" he shouted.

Glasscock and his wife, Lynne, who had brought the K-Y jelly for O'Donnell to swab in the well, watched as O'Donnell scooted out of the rescue shaft before taking his footing. He did so quietly, with a thin grin spread across his face.

In the moments following the rescue, the glory only ebbing slightly as national and international media made their way to airports to leave town, O'Donnell, Glasscock, and Forbes responded intermittently to requests for appearances and speaking engagements.

Already, though, Glasscock had stepped into another realm where attention poured onto him. O'Donnell stumbled into his fame at hour fifty of the rescue, when he broke down in an emotional rage over not being able to save the girl following his first attempt to reach her from the rescue tunnel.

Lilly quietly left the rescue site and found a hotel in Midland, coming away from the scene exhausted. He dusted off his clothes, put his mining hard hat on the floor with his boots, and sat on the edge of the hotel bed to call his wife at home. Throughout the rescue he had not talked to her, not wanting to bother her with the dramatic rescue or take his focus off the work. Rarely had he glanced up from his post near the well throughout the entire ordeal, noticing only briefly the growing throng of media.

Lilly reached for the hotel room phone, dialing his home number where Doris answered.

"You can't imagine what I've been doing," Lilly told his wife.

"What do you mean?" she asked. "We've all been watching you on TV."

CHAPTER 16

WHAT HAPPENS NEXT?

The Aftermath

THE COPS AND FIREFIGHTERS FORGOT ABOUT ROBERT O'Donnell.

The ambulance carried Baby Jessica and Cissy through the Midland streets to Midland Memorial Hospital a few miles away, but a glow of celebration remained behind at the scene of the rescue. Glasscock, looking relieved, draped an arm over Lynne and nuzzled his cheek on her head, pulling her in close. Volunteers and first responders slapped each other on the back, exchanged high-fives, and one by one they were plucked out of the sea of hard hats and oil field ball caps for interviews by media. But O'Donnell was nowhere to be found. He was still at the bottom of the rescue shaft.

"Where's O'Donnell?" Felice asked, looking around the shaft where crews continued slapping each other on the back.

The bedraggled firefighter-paramedic was picking up the gauze wrappers and trash at the bottom of the shaft, and it took a while before anyone realized the rescuer was nowhere in sight.

"Miller time," one rescuer hollered out, clearly satisfied with their accomplishments.

Roberts peeked over the edge of the shaft, planting a foot firmly in place on the rim of the hole. O'Donnell, still cleaning up the mess, looked up into the glowing circle above him. O'Donnell's heart raced as one of the men who had directed the drilling pointed a finger down into the rescue shaft, where the line moved and shook with O'Donnell attached to a harness.

Roberts moved back toward the rescue line to grab it alongside other hands that were guiding it along. Suddenly, O'Donnell burst out and lurched forward, standing chest to chest with Roberts. Glasscock and Lynne were only feet away, with others watching closely. A sudden applause erupted among whistles and cheers as O'Donnell began unbuckling his harness, his two-tone blue cap dirty and dusty. Roberts, standing close to O'Donnell, reached for O'Donnell, who grasped Roberts's arm and ducked his head in a partial hug.

"This rescue would have to be the most exciting time of my life and career," O'Donnell said.

Forbes did not hold back at the time either.

"The joy I felt at the time they were raising us up is the same joy that I had when I got married and when each of my children were born," he said.

The joy rippled in waves across the world, reverberating out from Midland and easing the hearts of those who had committed to watching the news coverage of the rescue. Charles Boler, the driller who had shown up at the scene with his crew of drillers early Thursday morning, sipped coffee at his breakfast table Saturday morning, pondering his participation in a truly remarkable event—one about which the world was talking.

It left the hardened West Texan with an itch to express something somewhere deep within him. Sitting over oily black coffee, Boler scratched out a poem.

> When you think you're down and out,
> The story of little Jessica
> Makes you think of what life is about.
>
> We drilled. We hammered.
> Everyone worked his heart out.
> In the end when she was lifted out,
> We had really seen what life was truly about.

"It felt good," Boler said of the words he wrote down in the moments he searched for some way to show how he had come to feel.

At the same time, workers were still at the well, capping it. Toby Partridge, the paramedic who had driven the ambulance to the hospital with the prized Jessica inside, planted a redbud tree by the capped well, surrounded by a bright colorful ring of chrysanthemums.

Willie Thames, one of the workers who had helped sharpen drill bits during the operation, used solder on the well's cap to write, "For Jessica 10-16-87 with love from all of us."

Most of the on-site media stayed for wrap-up pieces. Wunsch rambled on for a few minutes, staying on the air live as KMID producers rushed cameras to the hospital and figured out how they were going to get their live truck out of the parking lot that Tanner Drive had become. Wunsch babbled on almost incoherently, attempting through his tired stupor to wrap the fifty-eight-hour struggle into a tightly wound package.

Reporters struggled to find those who had pulled Jessica free.

Down inside that shaft, it had been O'Donnell who did the actual pulling—the one man who, in that last-ditch effort, tugged at her slowly, inch by inch, to free her. Forbes, though, who had emerged with the bundle swaddled in gauze, was the face of that final moment when she emerged. That caused confusion. Forbes and O'Donnell were interchangeable heroes. Forbes's bright grin filled television sets as he looked on at Jessica's amazing rebirth.

Reporters fell upon themselves, reaching out to Forbes to determine how he managed to free the baby. By the time the story could be pieced together, O'Donnell's swarm followed him from the hole through the alleyway.

"I consider every one of us who had anything to do with it as heroes from the chiefs all the way down to the volunteer workers that was bringing us food," he told a reporter, a hat sitting crookedly on his sopping wet head.

CHAPTER 16

O'Donnell could not see what was coming. As he and Forbes responded to media requests, momentum shifted to prioritizing their profiles as the faces of the rescue over those of the other volunteers and first responders. While O'Donnell and Forbes often would give credit to the broader group, they nonetheless served the media as proxies for the rescue effort. The interview for the *20/20* TV news magazine at MFD Station 6 the following week was where the two were prepared for their interview, with the station as a backdrop. The setup took the station and its vehicles out of service. Officials shut it down for the lights and camera crews, which were quite substantial for a production like *20/20*. Calls normally routed to that station would have to be assigned elsewhere if an emergency call came through. Forbes and O'Donnell took their seats on the bumper of an open ambulance for the interview.

The Midland firefighter-paramedics mostly looked on while the two smiled and quipped with jolly backslapping camaraderie. Though while O'Donnell's own shift handled his growing popularity in stride, others watched and judged.

"We were all there," one firefighter after another mouthed off. "We all did this, not just them."

O'Donnell embraced his fame. Robbie's mother stitched a blue-and-white cloth cover to a photo album, bordering the centerpiece with bundled navy-blue cloth with white hearts. In the center, stitched in more blue lettering on a clean white background, are the words "Our Hero."

Jealousy swelled in the ranks. And still, the calls kept coming from the media and others.

Although Forbes accepted interviews initially, he noticed the other firefighters getting resentful of the attention he and O'Donnell received. A year after the rescue, Forbes turned down interviews during the extensive anniversary coverage of the event.

Noticing O'Donnell's grip on the limelight, by contrast, Chief Roberts saw the two men evolve into their respective realities. Requests for interviews spilled into Roberts's office, and he doled them out to either Forbes or O'Donnell as they came in, though he laughed to himself, pondering the outcome.

"And I always knew that Steve wouldn't call. And Robert would," Roberts told The Learning Channel.

Resentment grew and then festered.

Forbes and O'Donnell traveled all over the country to the call of many—to EMS and fire houses for speeches, to Austin at the invitation of Texas Governor Bill Clements for an awards ceremony. The Veterans of Foreign Wars presented O'Donnell with their Firefighter of the Year Award for 1987.

On May 5, 1988, O'Donnell traveled to Charlotte, North Carolina, where the *Charlotte Observer* staff writer Tex O'Neill described O'Donnell as a "humble hero." Speaking before a group of 1,400 firefighters at the American Firehouse Exposition and Muster in the Charlotte Convention Center, O'Donnell said, "We do this every day. It just doesn't take fifty-eight hours and we don't get national attention. I don't think there is one person in here who wouldn't do the same thing."

Even as his words spilled out, adrenaline coursed heavily though his body. It stung his arms, and his spine straightened into a rod of rigid steel. He felt strong and tough, invincible.

Back at home, he was not looked upon in the same way. Few of his fellow firefighters held back their mockery, and sometimes, the pestering would drive him to tears.

"They're just stupid," O'Donnell told his wife, coming home from the station after bearing the brunt of his colleagues' building scorn. "I don't know what their deal is. They think I'm taking up all this media spotlight. Look, these reporters call me for a story. I give it."

Robbie felt her husband buckling under the pressure of the spotlight. If it was such a big deal for the world, if it was such a big story, then he would tell it for them, she related.

Celebrity status for O'Donnell and Forbes came in measurable increments. In August 1988, Forbes and O'Donnell—along with their wives—boarded a jet bound for Los Angeles. They were invited onto the game show *Third Degree*. The duo's appearance on the show forced celebrity guests to determine just how the two were related after the host Bert Convy delivered only a few clues.

Stepping out onto the set, the two appeared side by side as Convy rattled off the clues for the celebrity contestants, pinning the two as "true American heroes."

Such appearances became the norm. On the trips, the two men were treated like heroes, escorted around town to lunches and dinners and housed at first-class hotels neither could imagine staying in otherwise. The Los Angeles Fire Department escorted the two couples high above the city, granting a tour aboard their new helicopter. Their trip included scenic visits to see the *Spruce Goose*, the *Queen Mary*, Dodger Stadium, the Playboy Mansion, and the Hollywood Walk of Fame. The quartet dropped by Venice Beach—a couple of West Texas kids taking a walk along what O'Donnell could only describe as "nutzoid" to Crimmins in an interview about the *Third Degree* appearance.

"There were people singing for donations, one guy was juggling chain saws," O'Donnell explained to Crimmins after the trip. "We saw a one-man band and a Hare Krishna revival."

O'Donnell would only shake his head and smile. There was a world of difference between conservative West Texas and the whiz-boom-bang of Hollywood. It shocked and amazed him.

On May 9, 1988, O'Donnell and his wife traveled to Detroit for a presentation of the "Others" Award from the Salvation Army. On September 8, 1988, O'Donnell gave a slide presentation in Jekyll Island, Georgia, to a crowd of more than a thousand at the 1988 Emergency Medical Symposium.

Other trips did not carry the pizzazz of Hollywood, leaving O'Donnell little to talk about when he returned home. With each request to speak publicly about his rescue experience, O'Donnell gladly responded. In October 1988, O'Donnell spoke at a conference in Le Sueur, Minnesota. He was reimbursed $148 for the rental car he paid for out of his own pocket to get to the tiny town. The emergency workers who had brought O'Donnell in to speak had to make last-minute adjustments when it was revealed that O'Donnell was limited from certain uses of his voice and likeness due to a pending movie deal. Organizers of the trip had to arrange for his speech to not be recorded because producers were simultaneously working on the film about the rescue.

The attention and the media onslaught could not last forever. But he tried to stretch it out as long as he could.

"I knew he was talking about the stress and shit. The stress ain't what would get you," Ricky remembered his brother surmising. "It was the way today you're a hot story, and tomorrow if you ain't hot, they'll drop you like a hot rock."

CHAPTER 17

CHAPTER 17

OPRAH COMES TO TOWN

BEFORE RESCUERS COULD CART OFF EQUIPMENT FROM the scene, talk of a movie began filling the air. With a nonfiction plot line that read like a red-hot fiction adventure story, the prospect of Baby Jessica and her rescuers on the big screen seemed a given.

"The plight of the little girl trapped in an abandoned well in West Texas was a made-for-television story of the day," wrote an Associated Press reporter in Dallas, developing an angle on the vast attention paid to the real-life drama.

Other media sent out prophecies of what might be to come.

"The rescue would have been dramatic enough anywhere. But, played out in this struggling oil town with a grimy cast of hundreds, it had the tightly scripted feel of fiction. If ever a rescue effort took place at an appropriate time in an appropriate place it was here," wrote Peter Applebome, a special correspondent to the *New York Times* at the time of the rescue.

At about the same time as President Ronald Reagan and his wife, Nancy, were putting in calls to the McClure's hospital room, Oprah Winfrey's producers were already on the march, preparing a show on the rescue—to be staged in downtown Midland, the day after an elaborate parade to celebrate the rescue.

"In bringing our show to Midland, we hope to recapture that same sense of community as we give recognition to the bravery of Jessica, the perseverance of the McClures and the valiant efforts of the hundreds of volunteers and citizens who gave of their time, skills and prayers," Winfrey said in a written announcement of the show's airing from Midland.

The show was big for Midland. It was big for Oprah as well. Though she was on the rise, Oprah had yet to reach even close to the star power she would eventually come to enjoy. Having gone into national syndication only fourteen months before, the *Oprah Winfrey Show* had rocketed to the number one talk show in the country. She captured three Daytime Emmy awards in 1986, and by the time her crews began unloading trucks in Midland, the show aired in 180 cities, covering more than 130 markets. Oprah's show flickered across 98 percent of the television sets in the United States. The woman who knew what struggle and hardship was related to Midland and its overcoming of an impossible task.

Oprah's visit to the town capped the glory of the rescue but provoked—and perhaps catalyzed—the beginning of a swirling feud among rescuers over the national spotlight. While some three- to four-hundred people participated in the rescue, the media began narrowing the list of key players before Jessica's ambulance had pulled into Midland Memorial Hospital's driveway.

"They had selected a very few people to do the interviews. And a lot of people were left out," said Glasscock, who took a seat among the throngs of rescuers—separate from a select few who were placed on stage for the show. He watched as Lilly, O'Donnell, and Charles Boler took seats on the stage, fielding Oprah's interview questions.

Twelve hundred people packed Midland Centennial Plaza on Friday, October 30, 1987, to film the show. Producers had prepared the crowd beforehand, tipping them on how to look good on TV: do not chew gum, put purses under chairs, do not get up for a bathroom break, and for the love of God, put those homemade signs away. The crowd, prompted by an applause controller who told them when to start and stop their applause, was also shown how to applaud the host. Clearing the screen of distractions, producers told the audience they wanted the rescuers to express their true feelings and experiences to make the show a success.

"It's been told over and over again, and you're getting tired of it," a spokesperson who warmed up the crowd said. "When you sit down by yourself, you still think about it . . . the whole world watched you and wants to know how you felt."

When Oprah appeared, the crowd burst into applause as she announced, "America is alive in Midland."

In her insightful way, Winfrey may have been capable of understanding the effects of her show herself and its ability to boost men and women to fame.

"Do you feel like all of the eyes of the country are upon you?" she asked Dr. Sheldon Viney as he stood in a hospital wing at Dr. Charles Younger's side to give their assessment of Jessica's condition.

He replied, "We've felt like that for about the last two weeks very much."

She deduced, "I guess the *Oprah Winfrey Show* doesn't help the situation too much."

Afterward, Oprah's vans loaded their cargo, KMID-TV cameramen unclipped their cameras from their tripods, and the Midland Center was swept clean.

Although Oprah had staked her position, other big movie and TV producers had already put the story of the rescue on the radar with trips to Midland just days after the rescue's completion. By the end of October, just as Oprah's team left town, a slew of producers were finalizing proposals and meeting with rescuers. The day after the Oprah broadcast, Mary Alice Kier and Karen Danaher Dorr of Los Angeles's Ohlmeyer Communications Company (OCC) pinned down O'Donnell. Dorr, the OCC's vice president of Creative Affairs, discussed their vision for a drama, sharing O'Donnell's desire for a movie depicting the cooperative effort of the community to save Jessica.

Three days later, on November 3, 1987, Dorr continued courting O'Donnell, offering him $1,500 for the rights to his side of the story on a six-month option by a studio. Securing O'Donnell's rights would be a key in the story of the rescue, even though hundreds of people had participated in it. If a production company did not have the rights to the man who pulled her to safety, the possibility of their getting the movie deal would be slim. OCC's offer included another $2,000, opening an extended option period if a studio did not pick up the story within the first period. When a studio did pick up the project and OCC moved forward to produce the project, O'Donnell would receive another $20,000—minus the amount he had been fronted to secure his rights.

Formed in 1982 by Donald W. Ohlmeyer, OCC teamed up with Nabisco Brands, Inc., to link commercialization of the company to television programming. Ohlmeyer had similar arrangements with Nabisco for other made-for-TV movies, too. *Special Bulletin*, a TV drama he championed for NBC, won an Emmy in 1986 for Outstanding Television Drama Special. Ohlmeyer won recognition as producer and director for both ABC and NBC, and he gained attention for his efforts with ABC's *Monday Night Football*. He would eventually lead a turnaround for NBC with hits like *ER*, *Frasier*, and *Will & Grace*.

The New York CEO of OCC knew the game well. The OCC team had met with O'Donnell several times, peppering him with letters and contractual agreements that struck a tone of formality rather than gaudy but dubious promises of fame. OCC's letters and offer package likely caused O'Donnell's eyes to bulge. O'Donnell was a good ol' boy living in a middle-class home in West Midland, and the prospect of $20,000 in his pocket would have been alluring.

Other offers began flooding in.

The twenty-nine-year-old father of two who had dropped out of Midland High School in the late 1970s became the focus of contracts and movie deals with a labyrinth of ifs, ands, or buts. Packets came in the mail offering thousands of dollars to secure his interests in their options for production companies. On December 8, 1987, almost two months after plucking Jessica from the well, O'Donnell signed contracts to secure the services of LA attorney Sheldon G. Bardach, agreeing to pay Bardach 15 percent of the funds he would earn in the movie deal to oversee his interests. O'Donnell was juggling plenty of requests and issues.

In a handwritten note, he scribbled a reminder from MPD spokesperson Corporal White: "Jim White said do due diligence on whoever gets project." Other notes include brief assessments of calls from producers. "Theresa Saldono was not good experience," for example. Other notes included reminders to send résumés to various motorbike companies such as RaceTech in Pomona, California, and White Brothers in Garden Grove, California.

"I think the main problem is that we were inundated with stuff we didn't know much about," O'Donnell later told a reporter. "We were in over our heads."

It was never supposed to be like this. There should never have been emerging groups of competing parties and factions, and that's because on October 22, 1987, just six days after the rescue, Cissy and Chip McClure told the media they were not interested in selling their story for a movie or tabloid coverage.

"It's easy for me to turn down all that money," Cissy said. "It's Jessica's story, and if she wants to tell it when she gets older, she will. It's not our story. It's hers. We're not interested in exploiting her." According to a wire report from United Press International, Cissy said she had already been approached by screenwriters even while they were still in the hospital, a few days following the rescue. Iain Calder, the editor of the *National Enquirer*, confirmed that while his paper made an offer to the McClures for their story, the rumors of the $100,000 payment for the story were overstated.

During the jockeying for rights to the story, factions began to form. Midland mayor Carroll Thomas was concerned that the disagreements, which were becoming more toxic, would soon dismantle the goodwill the rescue had generated for his town. He appointed a blue-ribbon panel charged with mediating the dispute over movie rights, and he appointed Wayne Merritt, a local bank chair, as the group's leader, along with other notable members of the community.

By January 28, 1988, another offer would be on the table—this one from MGM/UA Television Productions in Culver City, California. MGM/UA's vice president of business affairs Susan E. Brooks knew the

CHAPTER 17

rescuers had developed the McClure Rescuers Association—a group made up of rescuers who would help select the movie's producer—to represent and speak for the group as a whole. It also gave producers fewer targets to wrangle for negotiations.

Before Christmas, a bloodthirsty competition developed among producers who had jetted into Midland, offering smiling promises and firm handshakes. Among them was Loehr Spivey of the Highland Communications Group of Van Nuys, California.

Loehr Spivey had stepped off a plane at Midland's International Airport in late October 1987, just a week after the rescue. The striking flat desert hit him with the full breathtaking force of a West Texas dust storm. Waiting for him at the terminal was Charlie Mendenhall, a Hilton bellman originally from Pittsburgh. Spivey listened as Mendenhall manhandled the conversation as though he had been born naturally to Southern hospitality, telling of his own arrival in the small West Texas city as his manifest destiny, and how he and his family thrived in their own small heaven.

Mendenhall steered the Hilton courtesy van into a circle drive near the entrance of the hotel—the closest to a five-star Midland had.

"I guess you might have heard about that thing that happened out here with little Jessica McClure," Mendenhall queried Spivey with a hint of a West Texas accent creeping in and not knowing he was talking to a producer with more than a few made-for-TV credentials under his belt.

Mendenhall walked Spivey up to his room, towing a luggage cart.

"So, what do you do out in LA?" Mendenhall asked.

There was no way for the bellman to know the passenger he picked up at their airport was one of two major players in the battle to get dibs on Jessica's movie.

"To lose the opportunity to produce this story would be almost as tragic for me as losing Jessica would have been for the community," Spivey wrote in a November 1987 proposal to the rescue association representatives. "I have met and spent time with a substantial number of folks and have learned a lot about the community and its character."

But, to many, meeting him left them with a bad feeling. Some of Spivey's talk was a little too slick for their liking.

"We should have tarred and feathered him and run him out of town on a rail," said Glasscock later. Spivey constantly called the rescuers, O'Donnell among them, getting calls through to him at the fire station. Still, Spivey avoided laying out specific details of the story he wanted to tell, trumpeting a vague but spirited vision for his drama.

"Through this project we potentially have the instrument with which to reach out and touch many, many millions of people in this country and the world in a totally positive manner which will, I believe, leave tears of pride and joy in the eyes of all who see our film," Spivey wrote, proposing an editorial committee made of Czech, Roberts, *Reporter-Telegram* publisher Charles Spence. and others to oversee drafts of the script.

In February 1988, Spivey tried to offset public concern about Midland's portrayal in the movie, assuring accuracy and public access to dailies. He could crack only one member of O'Donnell's group and kept finding resistance as other offers poured in and as others involved weighed in against Spivey. He began attempting to scoop up what rights he could.

On February 10, 1988, Kragg Robinson, the rathole rig operator who had pulled his drill bit while digging pilot holes for massive overpass support columns on Loop 250, retracted his membership from the McClure Rescuers Association.

"It was represented to me prior to joining that the association would continue to invite many additional members and would be truly representative of the rescuers in the community," Robinson wrote in a letter to association members. "This has not been done. Accordingly, I cannot in good conscience continue to be a member of this association."

Spivey had been the one encouraging O'Donnell to form the association, O'Donnell told reporters. O'Donnell said forming the association was intended as an effort to corral rescuers' stories into one reasonable approach to the movie. But it was a pained and fractious effort from the start.

"Our goal is to have one [a movie] made that depicts the thing as it actually happened," O'Donnell told the *Reporter-Telegram*'s assistant city editor Rick Brown. "We don't want a Hollywood version."

CHAPTER 17

But the association stumbled. Some splintered and formed their own associations. Allegiances shifted. Membership swelled and fell off.

Czech and Roberts were on board as a first responder group together with Glasscock, O'Donnell, and others. But at the same time, other officers declined.

Sergeant Jeff Haile, a media go-between at the rescue scene, opted out of the movie deal.

"I thought it was unethical," he later told the *Odessa American* newspaper. "We had a lot of people involved in the background who played just as big a part as anyone who went down in the hole."

The public-relations rainbow, which formed above the city in the days after the rescue, began to fade. National media began reporting the clashes among rescuers.

Lisa Belkin pulled into Midland at the height of the squabbling, after having covered the rescue from the scene as it unfolded on the pages of the *New York Times* through their Houston Bureau. Belkin noted O'Donnell's easy manner around reporters, citing his concise sentences as he constructed them in sound bites.

"I hate to see it split the rescuers like this," he told her for her coverage. "But our story is the real story—we were the major players. I want the best movie. One that my great-great-grandkids can see."

The developments irked the same Americans who had gushed with pride as Baby Jessica was brought out of the well in Forbes's arms. Denise E. Gilvey, of Santa Ana, California, voiced her own disdain in March 1988, sending a letter to O'Donnell and Robinson after reading Belkin's piece when the Associated Press wire ran with news of the clash. A copy of the letter ended up in the hands of the editor at the *Orange County Register*.

"In these days of 'no news is good news,' you all bonded together to help another human being. We watched the TV and listened to the radio, eagerly, hoping that Jessica would be rescued alive. It was the type of tale that brings a lump to your throat," she typed, taking the time later to underline key words to hammer her point home. The letter, at first, sounded like a thousand others O'Donnell kept—mailed to him from around the world. "But like everything else, poor little Jessica's story is being commercialized," Gilvey wrote. "You are all now making a mockery of yourselves in front of the whole nation. You're turning a sad story with a happy ending into profit for yourselves. If you take a good look at the current situation, you would probably be just as disgusted as I am."

"The news people have besieged most of us," a March 29, 1988, letter to the association reads. "To avoid contradictory statements by different members of the Association, it is recommended that no one make statements to the press or TV of any nature whatsoever at this time. However, it is believed that we should be cool and steadfast in our desire to get the best producer possible and fighting in public with the Spivey group will not help Midland."

The one movie deal that seemed to be gelling was that with Interscope Communications, a Los Angeles company with branding by Campbell Soup Company. In January 1988, they kicked up their total offer package from $100,000 to $200,000.

"Robert, they sweetened the pot to $200,000," a memo reads.

The March 10, 1988, offer from Interscope seemed like the one for O'Donnell and his group.

"With respect to the Jessica McClure story, there is no indication in the letter as to how much or how Campbell Soup will help in the making of the movie, but the involvement of a large domestic corporation adds comfort to the McClure Rescue Association that I represent," the association spokesperson wrote.

Darrell Smith, the McClures' attorney who also had a seat on the mayor's blue-ribbon board, held most of the cards, as the McClures would have the final say no matter what the rescuers' fight decided.

"Their biggest concern is they are grateful to the community and that the story be told properly if it is told at all," Smith told the media.

A *Reporter-Telegram* editorial expressed a desire for the end of the feuding with what appeared to be a final deal that would satisfy first responders, volunteers, and the McClure family. "But this puts the biggest hurdle behind us and allows the sensationalized coverage to end," the editorial read.

CHAPTER 17

Highland and Interscope gave their presentations to the blue-ribbon panel April 7, 1988, each putting forward their best proposal for the development of Jessica's rescue movie. In the end, only one winner emerged. And on April 22, 1988, Mayor Thomas's blue-ribbon committee announced its decision to move forward with Interscope. In the announcement, the panel made it clear that the rescuer group that included O'Donnell and ten other municipal employees had been disbanded,

Almost a year later, on February 27, 1989, O'Donnell penned his signature in wavy curls to a contract, optioning the association's interests with Interscope for six months for a down payment of $13,334.

Glasscock said the factions within the police and fire departments became even more divided. With the new deal, Interscope could seek out their own key players, defining for themselves who would be portrayed in the movie. "Even in our own house, there were people showing animosity. You could tell there was people upset about it," Glasscock said, and he blamed Spivey for the ongoing discord. "[Spivey] should have been run out of town on a rail," he said, summarizing his frustrations.

Whereas before his interest rested with the team that promised to deliver the most accurate story, Glasscock quickly sided with Interscope in all matters pertaining to the movie's development in order to get a seat at the table.

"Now I was on the winning team," Glasscock said later. "I was on the Interscope team."

On March 15, 1989, Glasscock and sixteen others were given $7,843.17 each through a ClayDesta National Bank account. The seventeen rescuers divvied up $133,334. The McClures were given $66,666 for their stories—one-third of the total $200,000 package. The total production aimed for a cost between $2.6 million and $2.7 million.

The announcement of the actual details that had come eleven months earlier with an official press conference at Midland City Hall the morning of Friday, April 22, 1988, was supposed to have been an official farewell to the strife. Once and for all, there would be clarity with a plan to move ahead in unity and agreement.

O'Donnell sat front and center in a light-colored suit.

"The two groups from the start have had the best interests of the community at heart," said Mayor Thomas. As O'Donnell and Corporal White left the press conference, a Fox network reporter moved for a comment. Neither said a word.

"I've been working since Monday, getting statements from the volunteer group, guys," the reporter begged.

White and O'Donnell huddled together for a moment, trying to determine whether they should or could respond, still sticking to the suggestion from Martin Allday, the group's lawyer, to stay quiet.

"No, we don't have anything else to say," O'Donnell said.

The Fox reporter quipped back, complaining about their decision to stay mum.

"Now that we're in agreement, y'all are trying to get us to fight," White fired back, walking away with O'Donnell. The blue-ribbon panel and the first responders may have thought they agreed, but the last reporters had heard from Spivey before the announcement indicated that he still possessed enough of the volunteers' rights to seek out a studio deal for a competing movie. Late on Friday, word of Interscope's deal for a movie still grated on Spivey, who fired off his own announcement of plans to file suit.

"Defendants in the suit may include Interscope Communications, Inc., the company that a panel of prominent Midlanders Friday recommended produce the television movie, and Martin Allday," *Reporter-Telegram* assistant city editor Rick Brown reported for Saturday afternoon's edition.

Patricia Clifford, a producer for Interscope's effort, shot back, "It sounds kind of ludicrous, honestly."

Interscope's producer John Kander moved forward, enlisting David Eyre to write the teleplay. Before Eyre could even put pen to paper, ABC called with an air date—May 21, 1989. Broadcast before the throngs of Sunday night viewers, *Everybody's Baby: The Rescue of Jessica McClure* would bring alive what had brought the world to sit on the edge of its seat for fifty-eight hours.

CHAPTER 17

Getting past the fight over the producers was not the end of it for Midlanders. Skeptical and apprehensive about Hollywood's adaptation of Midland, Texas, locals professed ardent concern for what might come of a movie about the Tall City. Could Tinseltown make the good people of West Texas look like backwoods hicks? Opening themselves up to investigative efforts to pry into the story and evoke what went on behind the scenes of the rescue could skew what took place—the way the world saw it live on their televisions.

"If it's going to be in Midland, let's do a good job for Midland. If it's not, let's not even bother with it. We've got enough positive out of that situation that we don't need a movie," Czech told a reporter.

Chief Roberts, seeing what the glow of Hollywood offered, maintained calm, trying to remain unfazed by the prospect of the developing movie.

"You have to remember I have a fire department to run, and that's a lot more important than a lot of things," said Roberts afterward. "What's $10,000? What's $20,000 if it's going to portray us in a bad light?"

O'Donnell wanted to make sure they would not. He knew the teleplay's author was conducting research even as the fighting had taken place over the last several months.

Eyre heard a knock on his hotel room door as he went over notes gathered in Midland from a number of interviews with police, rescuers, and the McClures. He had been doing research for the screenplay. On the other side of the door was O'Donnell, his arms folded and his hip cocked. O'Donnell spoke directly, asking if Eyre was the Hollywood writer picked to tell the Baby Jessica drama.

"He was ready for combat. He kind of had a chip on his shoulder," Eyre remembered. "I think he was worried about this guy coming from Hollywood to write this story."

Eyre invited O'Donnell into the room, O'Donnell carrying with him a crooked brow and the caution of a newly indoctrinated hero. Eyre broke the ice and chipped away at the paramedic's hardened facade.

"He warmed up to me pretty quick, because I'm a pretty easygoing guy," said Eyre.

Although O'Donnell was the only one to pay Eyre a visit and size up the situation, Eyre also had work to do of his own. He wasted a two-hour drive to Carlsbad, New Mexico, to meet with Dave Lilly, who let him know federal regulations would not allow him to benefit from the movie. The work ahead for Eyre was as difficult as that of producers. With enough news coverage to fill a library, Eyre could easily outline the event, but dividing the events into real-person drama for each of the major players meant one-on-one interviews and tearing people apart, piece by piece, to develop characters while telling the story of Midland.

"There was a total commitment of everyone to put this together as honestly as possible," said Eyre.

For the most part, he did a good enough job, said many Midlanders. Few expressed much irritation at his efforts.

On Sunday, May 21, 1989, many in Midland gathered around their television sets—some at watch parties and some with their own families in their own homes. O'Donnell, who had gathered his family around the set, fidgeting and popping off, was telling everyone to watch for his part. He actually had two parts—the one where actor Whip Hubley played O'Donnell as the rescuer who crawled desperately into the rescue tunnel and the one where O'Donnell himself acted in the film.

Hubley, who had starred as jet pilot Hollywood in the movie *Top Gun*, did little to adapt O'Donnell's twang, seeming to play the part as an extension of his own persona.

Glasscock and his wife met up with friends at a watch party in a small clubhouse at the Saddle Club Apartments in North Midland, waiting to see actor Walter Olkewicz portray the police officer who went to the rescue scene just to see what all the fuss was about. Olkewicz, who eventually also starred on the television show *Grace Under Fire* as the lovable but hapless "Dougie," had come to Midland and had ridden along with Glasscock on a shift. The two took Polaroid photographs together; Olkewicz even invited Glasscock out to his home in Burbank, California. He wrote his phone number and Burbank address on the back of one of Glasscock's business cards before he left town. Tapping a keg of ice-cold beer, the twenty-five people gathered at the party dealt out a wisecrack here and a giggle there as cheesy lines attempted to capture the culture of Midland in a two-hour movie.

"I thought parts of it were a little corny," O'Donnell told the *Reporter-Telegram*'s Brian Pearson. "No matter what they did, I probably wouldn't be satisfied."

The McClures watched the movie by themselves in their home with only mild complaints—the main one from Chip for his portrayal (by actor Will Oldham) as too "scruffy."

As they watched, O'Donnell continued shushing guests at his house, waiting for his moment on the screen playing a role in the movie.

"Hold on. It's coming," he insisted.

It never came. Instead, his part had fallen on the floor of some editing room in California—the pieces of film brushed away. No one had told him he had been cut from the final version.

The film climaxed with Hubley tugging on Jessica until she came free from the hole and ended with Beau Bridges (playing Chief Czech) and Pat Hingle (playing Chief Roberts) guessing that the probability of walking on water was indeed possible.

Of the characterization of the two as Christlike figures, Robbie O'Donnell said, "That sounds about right."

CHAPTER 18

A HOSPITAL UNDER SIEGE

DOCTORS AND NURSES AT MIDLAND MEMORIAL HOSPITAL had begun preparing for Jessica McClure's arrival two days before she finally was rescued. Hospital spokesperson Laurie Johnson handed out press packets, doctors considered treatment possibilities, and security cordoned off a section of the hospital for the throngs of media—some of them already camped out preparing advance stories.

At 8:03 p.m. on Friday, October 16, 1987, paramedic Toby Partridge turned his ambulance around the corner to the Emergency Room entrance, shooting up a wide corridor of concrete to the doors as applause and cheers broke out among the staff. The fluorescent glow of the entryway lit the area where the workers' containment burst into raucous clamor. Blaring car horns filled the night as drivers tuned into on-air radio reports and celebrated in the only way they knew how. The sound of their celebration poured across the hospital driveway.

Dr. Debbie Reese, the pediatrician assigned to treat Jessica, sprang from the ambulance; ER doctors were waiting at the ambulance's tailboard. Cissy climbed from the front seat of the ambulance, rushing for the hospital doors as the physician team swarmed Jessica, her eyes as wide as saucers, her skin pale and her expression confused. The concern became what lay beneath the swath of bandages.

And doctors knew every move they made would face an entire world of scrutiny—from the media, from the public, and from other doctors. Calls came in from physicians, suggesting a particular course of treatment. Other calls suggested home remedies from snake oil potions to massages.

CHAPTER 18

Outside the hospital, the pandemonium grew.

Reporter-Telegram reporter Julie Hillrichs turned around in time to see a car stuffed with teenage girls yell out, "We love you, Jessica!" The car was part of a jammed procession building at the southern end of Midland Memorial Hospital. The buildup of vehicles and their honking horns and their occupants in utter excitement clogged the intersection of Andrews Highway and Illinois Avenue near the entrance of the hospital. Traffic was jammed for more than an hour.

Security guards stationed at each entrance to the hospital limited access to friends, family, and media. Meanwhile, teddy bears—mostly yellow and red Winnie-the-Pooh stuffed bears—began piling up in the hospital lobby. Keshia Knight, a childhood star of the situation comedy *The Cosby Show*, sent a teddy bear named "Theo" to Jessica in care of the hospital.

Kimberly Modisett, the hospital's volunteer coordinator, expected the arrival of another three hundred bears for the hospital's gift shop by Saturday morning. Robyn McGraw, a six-year-old Midland girl, dropped off her own construction paper card, scribbling a message in colorful wax.

"Dear Jessica, we are still praying for you," McGraw had written.

Pink ladies, the volunteer candy-stripers who answered phones and ran errands, installed four phones at their front desk as calls increased to approximately fifty per hour.

Abra Kadabra, a gift and flower shop across the street from the hospital, attributed half of their sales that Friday and Saturday to people buying gifts for Baby Jessica. Co-owner Ferrell Powell watched in awe as people continued filing into the store, picking out one gift after another.

"You can tell some of these people don't have the money to do it, but they're doing it anyway," he said.

Sharon Boyd, Powell's business partner, dressed in a clown suit to deliver a bouquet to the hospital in anticipation of Baby Jessica's arrival that Friday. A five-foot box from Federal Express, the company that had arranged for the delivery of the hydro drill from Houston, contained a giant Winnie-the-Pooh bear. Marked on the box were messages from all over the United States as handlers looked at the packages' destination and penned their notes.

> FedEx in Midland loves you.
> Lubbock loves you.
> Midland Airfield loves you.
> Memphis FedEx loves you.

Money had been pouring in since news of the rescue began, prompting officials at KMID-TV to establish a special bank account at Texas Commerce Bank to handle the flood of donations to Jessica and her family. Others at the scene had also donated money to the family.

Midland County Sheriff Gary Painter put a wad of cash in Chip's hand during the rescue, telling him it was from someone at the scene who did not know how else to get it to Chip.

"I want you to count this right now," Painter said as Chip unfolded one hundred dollars from the wad, gawking in amazement.

Another man at the scene handed over the change he had in his pocket. Chip said envelopes would come with a simple note inside along with a check, some of them for twenty dollars, telling the couple to have dinner out or use the money for medical bills. Thousands of dollars poured in that way before an unknown donor forked over a large sum to foot the hospital bill.

At the hospital, the envelopes stuffed with money and checks filled the lobby along with the hundreds of bouquets and stuffed animals.

Despite the fanfare, Jessica's condition remained as tenuous as it had just hours before, when she was still in the well. For two and a half days, her right foot had been angled up tight against her body. Blood had stopped flowing through the leg long before her rescue, though she had no broken bones or internal injuries. Her blue toes lacked the blood to turn the skin the same pinkish color as before.

A large open wound to her forehead was the most visible sign of the young girl's trauma. Her head had been pressed against the well with

CHAPTER 18

a great degree of pressure—like pressing someone's forehead in a constant forceful push into a baseball bat wrapped in gritty sandpaper—for fifty-eight hours.

In an effort to force oxygen into Jessica's bloodstream, doctors placed her into a hyperbaric chamber at 9:25 p.m. She stayed there until 10:55 p.m., alone again in a small tight space cut off from everyone.

She was four pounds lighter than when she fell into the well; the first nourishment she consumed, aside from an intravenous solution, was an orange popsicle. She went into the well at twenty-one pounds; she emerged at seventeen. Although her toes began to redden after the hyperbaric treatment, doctors still expressed concern about the possible loss of the foot. Dr. Charles Younger prepared for a fasciotomy—a procedure in which a series of incisions are made to relieve swelling inside the muscle. The swelling could have blocked circulation, further hindering the foot from improving.

Many Midlanders—and the rest of the world—waited throughout the day Saturday and Sunday, still plugged into news outlets for the latest updates on Jessica's condition, though the suspense over whether the girl would live or die was alleviated late Friday night, when hospital staff announced Jessica's condition as "serious but stable." On Saturday, Jessica rested for several hours in her mother's lap between 90-minute sessions in the hyperbaric chamber.

In the meantime, preparations for celebrations in Midland went into full planning mode. Midland political confidant Bobby Holt announced Vice President George Bush's plans to stop in Midland after his last leg through Houston announcing his presidential bid. Bush's trip across the nation the previous week was part of his bid to succeed President Ronald Reagan. Holt had organized a last-minute reception for Bush at the Midland Center following a review of Jessica's condition at the hospital. "Look, you gotta go to Houston. You gotta go to Midland," Holt had urged Bush.

President Reagan and the First Lady themselves kept a close eye on the rescue. Now, however, the elation of a hopeful nation was tempered by a contracting stock market—signaling one of the largest single-day drops in the market's history. Even as the news of Jessica's rescue was being pressed into fibered gray newspapers across the country, other headlines painted an ominous picture of what was going on at the New York Stock Exchange just hours before. "Jessica's Out!" an above-the-fold headline screamed from the *Austin American-Statesman*, accompanying a photo of paramedic Bill Queen, who had carried Jessica from the well's edge after the handoff from Steve Forbes, who had been at the bottom of the well with O'Donnell. For a reader, it was easier to look at the wide, frightened eyes of Jessica McClure than to examine the headline in the next column over: "Wall Street Dazed by 108-Point Drop." The news foreshadowed Monday's market collapse that would shock the nation's economy.

The Reagans phoned the McClures' room. The call came at 11 a.m. on Sunday from Nancy Reagan's own hospital room, where she was recovering after having undergone treatment for a suspicious lump in her left breast. Though the biopsy's scheduling for 6:30 that Saturday morning had meant an early Friday night for the First Lady, she claimed later to have held off preparing for surgery until she found out whether Jessica was saved.

Cissy asked the First Lady how she felt after her surgery.

"I'm fine. How are you?" Nancy replied from her room at Bethesda.

Reagan himself had to interrupt the two mothers, reminding them of his presence. "I think you must be aware by now that everybody in America became godfathers and godmothers when this was going on," he said.

Later, on Sunday, Chip hurried around town running errands before returning to the hospital. When he noticed well-built men in suits and sunglasses looking alert throughout the hospital, he connected them to the shining black limousines and the color-coordinated Secret Service vans nearby.

Vice President Bush and his wife, Barbara, were already meeting with Cissy and Jessica. The Bushes looked on by the side of the crib where Jessica slept. Bobby Holt, also in the room at the time, watched quietly—and, as usual, stayed out of the Bushes' spotlight—while the vice

CHAPTER 18

president's official photographer snapped several shots. For fifteen minutes, the next president of the United States—who had once called this sleepy town home, the place where he had built his first business, started his family in their first home, and developed lifelong friendships—spoke with two kids: the father, a job-to-job house painter and roughneck and the mother, a seventeen-year-old daycare babysitter who had spent the past two days in tatters, waiting to learn the fate of her child.

For Jessica's doctors, who continued giving statements to reporters during press conferences, the spotlight was enjoyable but daunting. Dr. Younger worked to maintain perspective on the situation. Dr. Younger enjoyed being the center of attention for a time, but he said once was enough. All the doctors had practices. Other patients required their attention as well. Dr. Viney, the vascular surgeon, had a car wreck victim whose spleen had been ruptured.

For Jessica, though, the care was free. Viney volunteered his own services, which are estimated to have cost between five and ten thousand dollars.

"The guys who got her out of that hole didn't charge her," he said.

Younger did the same, unprompted and unasked.

In the end, a local anonymous donor paid for the hospital's $50,000 treatment, along with a donation for a hospital foundation.

Monday, October 19, 1987. Dr. Younger, an orthopedic surgeon, removed dead tissue from Jessica's foot at 11 a.m. in an hour-long procedure. Using a dye, the doctors detected exactly where blood could and could not flow. Only portions of the foot would take the dye. The color would not diffuse completely through the appendage.

Dr. Viney used a Doppler laser to examine the microscopic blood flow through the foot's skin during Younger's procedure. The capillaries throughout Jessica's big toe and along the outside instep saw the most damage. Dr. Viney and Dr. Younger, though, remained optimistic, knowing the skin would grow new capillaries to support blood flow.

A HOSPITAL UNDER SIEGE

The more bothersome injury—mainly due to its blatant appearance—was Jessica's forehead wound. The area appeared beet red and aggravated. Plastic surgeon Terry Tubb devised a plan to insert inflatable silicone bubbles beneath the skin on Jessica's forehead. In a news conference, Tubb displayed the device and explained its use.

"This is a large one," he said, holding up an opaque plastic device. "It has a remote valve. It is made of silicone. Inject salt water in this valve and fill up this reservoir, which stretches the skin. These are used for a constellation of other deformities, namely recon of breasts after breast cancer, removal of burns, skin grafts, tattoos, et cetera."

The cross-examination from the reporters left Dr. Tubb feeling a bit burned.

"We would have been chastised if she came out on a peg leg. The news media were very predatory at our news conferences," Tubb told *Reporter-Telegram* columnist and reporter Ed Todd in 2012, noting the questioning from media and what he described as a negative slant. "'Do you guys have the right stuff? Can you handle this? Why don't you transfer her to Harvard [Medical School's teaching hospital] or to Johns Hopkins [hospital]?' We had to stay real positive, because they were probing real deeply."

The skin that would have to be grafted was donated from Jessica's own hip. The bubbles would raise and stretch the skin over the entire wound, resulting in a small scar to the forehead. Although Tubb initially thought he would be able to get right in to perform the surgery, he would have to wait for more healing time. Younger would also have to wait to perform the skin graft to the areas where the fasciotomy had been performed.

"I'm going to slow down, take my time, let Mother Nature take its course," Tubb said. "I want the best condition possible."

The constant attention from a fully ramped-up hospital staff in the circular glassed-in critical care area of the hospital meant Jessica's wounds faced repeated exposure as the young girl's body worked to heal and staff checked and re-checked bandages and wounds. As six doctors checked throughout each day, the concern shifted.

"Children and babies like Jessica have great recuperative powers," said Viney. "The trouble with doctors and nurses is that they are getting overzealous."

Looking at the foot, Tubb's assessment emphasized the likelihood of her losing it completely. Doctors bandaged the foot and waited until Wednesday, October 21, to view the progress of the healing—while continuing to treat Jessica with several hyperbaric chamber exposures each day to force more oxygen into the blood. Still left hanging in the balance, Jessica's foot appeared too damaged to survive. On Friday, October 23, Younger removed more dead tissue from the foot, and though circulation improved, the right big toe did not look good.

"Physicians reported a definite probability that she will lose the tips of her little and big toes," hospital officials announced after Younger's exam.

Still, even as these announcements were made, Midland clung onto the limelight. Using terms like the medical staff's "definite probability" accompanied the efforts to show only the excitement of the rescue and recovery. Letting go of the rescue would have meant letting go of something that made West Texans feel good for the first time in a long time. Midlanders' efforts to use the rescue in such a way would continue for quite some time—for some of the rescue administrators, it would continue for years. To tamp down the negative aspects on the horizon, only a portion of the straight facts would filter through Midland's news reporting—none of it with the depth exercised during the rescue.

On October 26, doctors decreased the number of Jessica's hyperbaric treatments to two per day. The likelihood of removing toes appeared imminent. Tubb could not begin plastic surgery using the rapid skin expansion technique until November 4. By then, however, the celebration of the girl's rescue had filled and emptied the streets. The event had evoked a feeling of hope and opportunity. Jessica perpetuated the philosophies of those Midland conservatives who claimed the land catalyzed possibility for all those determined enough to achieve success for themselves and their families.

While the glory of those days after the rescue passed on, Jessica's health gradually improved. On November 19, Younger set a walking cast on Jessica's right foot. A few hours later, after more than a month under twenty-four-hour care by physicians and nurses, Jessica hobbled out of the hospital. While most of her foot was able to heal, saving Jessica's right little toe became impossible. In a surgery that ended at noon on November 4, doctors removed the toe as well as additional dead skin on the foot.

"Jessica's foot survived by a whisper," Tubb told *Odessa American* reporter Laura Ludewell a year after the rescue. Six operations, thirty-two days in the hospital, brand-new surgical techniques, and a world of prayers—all of it for one little girl who still suffered despite her salvation from the well.

CHAPTER 19

"I KNOW HOW I'M GONNA DO IT"

ROBBIE O'DONNELL REMEMBERS HER HUSBAND COMING home from a particularly grueling shift at MFD Station 2. Still living in their tiny house on Anetta Drive—where O'Donnell would park his car in front of the driveway so his kids could play without drifting out into the busy street—he looked particularly shaken.

That day, an 80-year-old man had gone into his backyard where he had penned a note of apology. He had placed the note on a fence post beneath his false teeth and eyeglasses. And then the frail, elderly man blew a hole in his head with a gun.

"Man, if I decide it's time to go, I know how I'm gonna do it," O'Donnell told Robbie. She would think little of the comment at the time—though it would eventually ring ominously in her mind several years later, only to stay there and fester over the course of her life. After his death, she would look back at so many of the things O'Donnell had said and done. His actions all meant so much more in retrospect. The words could have—and should have—struck an alarm, especially coming from a man who dealt with death and dying every day.

The same Christmas Eve that Ricky rushed to Huntsville to bring his sedated brother home to Midland, O'Donnell had called Robbie. Celebrating with her children at her parents' house—which was their tradition—Robbie heard the voice on the other end of the line driveling under the influence.

Robbie knew what the effects of the contents of O'Donnell's black duffel bag could be, and she heard on the phone that the pills had been consumed. Ricky, aware of his brother's knowledge of medicine as a paramedic, assumed O'Donnell knew the combination of pills needed to thwart his migraine. Robbie could do little but watch when O'Donnell claimed headaches and insomnia necessitated medication. She knew that without immediate relief, O'Donnell would treat himself to more and more.

She imagined it so clearly after having seen it all up close. Since their split in 1991, dealing with his stupors had become a thing of the past. Still, she could not help but think of those moments when she witnessed his addiction and confusion. Accompanying any task, the pills fell right into his hand.

Time to eat. Take pills.

Time to sleep. Take pills.

Time to go to work. Take pills.

He would ask her to retrieve some of the medication, and Robbie would fumble through the bag, sorting her way through pill bottles and stacks of unfilled prescriptions from doctors in different towns. She found Dr. Vogel's prescriptions. She found another from a doctor in Brownfield, and another from a doctor in Dallas.

Clever, she thought.

Among them, Halcion for anxiety, Prozac for depression, Fiorinal #3 for sleep, and dihydroergotamine (DHE) for headaches.

The last straw had come well before this Christmas holiday. She had given herself time to get over her husband's many flaws. Though it came at a price, she realized the split was necessary for both her own and the boys' safety.

She could not have another night like the one she had in April 1991, when she awoke to the rumbling drone of her husband's snore, slightly in awe of his deep sleep. Something, though, was wrong. The lights in the house were on—all of them. As she went to turn them off, a breeze blew through the back door, which stood wide open. Closing it, clicking light switches off along the way, she swatted at the insects fluttering all around her as they tried desperately to reach the warming glow of a burning light bulb. She then noticed the door to the garage standing open as well. Leaning in to turn off that light and close the door, she noticed the garage door, too, stood wide open.

She began to fume, thinking of her two young boys asleep in their rooms—and their father asleep in his and stoned out of his mind.

"This is it. I'm not living like this anymore," she said aloud to herself, moving through the house to double-check every inch.

Swatting at the flying, flapping insects, Robbie stormed back into her bedroom where O'Donnell continued his restive sleep. The snores bellowed from his thin, tight chest.

"Robert, you wake up," she growled. "Wake up, Robert."

The snores continued without a bit of change.

"Wake up, I said, wake up," she called again—this time raising her volume and standing on the water mattress as it sloshed around, almost throwing Robbie to the ground. O'Donnell still did not budge.

Her frustration morphed into anger, building enough to evolve into fright. She began to scream.

"You wake up, Robert!" she yelled at the top of her lungs, shaking and rocking the bed as she straddled O'Donnell.

Still nothing.

It took several more minutes of ranting before O'Donnell finally fluttered his swollen tired eyes.

"Get out. You get out of my house," Robbie sternly ordered, her baby-blue eyes wide in their sockets. She did not care where he went. She still does not know where he went.

Stoned and half asleep, O'Donnell puttered away from his home in his red Jeep, finding one place or another to crash.

"I was done," Robbie said.

The next month, Robbie filed for divorce.

At the fire department, work was not going much better.

On April 19, 1995, just hours after the reporting of the Murrah building bombing started, O'Donnell called Vaughn Donaldson. One of the only people to understand what O'Donnell was experiencing, Donaldson had become close with O'Donnell in the early 1990s while working with him at MFD's Station 6, the nearest station to the old Jessica McClure rescue site.

Despite his friendship, even Donaldson was also one of the firefighters who needled O'Donnell about his fame. There was quite a collection of them. Some claim remembering to have seen O'Donnell opening hero letters from admirers, emptying the envelopes and separating the contents into stacks—letters in one pile, cash in one pile, and checks in another.

Others who had been his friends now shunned O'Donnell or skirted him to avoid O'Donnell's talk of the rescue, like the high school Harry who cannot avoid reliving the saga of his greatest city championship football victory.

Being a friend with O'Donnell might also have become a liability. At least one firefighter recalls a MFD captain shooting O'Donnell full of lidocaine—a drug doctors use to numb the body.

CHAPTER 20

BOYS WILL BE BOYS

Colleagues Turn on O'Donnell amid the Spotlight

> "If you're talking about post traumatic, I don't have a whole lot I could tell you about that, because I wasn't traumatized, and I didn't see anyone else traumatized. The only one was O'Donnell, and he was traumatized before he even started."
> —Midland Police Chief Richard Czech, March 7, 2005

THEY CALLED HIM "ROBODONNELL." *ROBOCOP* HAD JUST been released.

In fire stations around the world, grown men exercise the playful banter and tomfoolery of children. Firefighters are notorious for pulling pranks, smart remarks, and jokes on fellow firefighters. It is almost a form of emotional attachment and bonding. It is a rite of passage.

Testing the stability and spine of a new firefighter started on a rookie's first day at the station—especially when the station was Central at 1500 West Wall Street, which also housed the main headquarters for the department. After the rookie was shown his place on the engine, he would be ushered inside while other firefighters would stuff his boots with wadded-up newspaper. They would turn the boots backward in the rookie's bunker pants—which were otherwise designed for the firefighter to step into before raising them by the suspenders to place around the shoulders. Using wire and string, the conniving firefighting veterans would tie the sleeves of the rookie's bunker coat closed.

CHAPTER 20

And then the alarm would sound, filling the engine bay with the voice of a dispatcher's command to respond to an explosion with fire. The firefighters would sprint to their bunker gear and emerge fully enveloped in the yellow suits in a single fluid motion.

"Respond to fire with explosion at 1500 West Wall Street," the dispatcher's voice would call out, backing the captain's shouts at the rookie struggling to get his feet in his boots.

"Let's go, rookie! There's gonna be bodies everywhere! Let's go!" the captain would yell. "We're gonna fire your ass on your first day if you don't move it. Now let's go!"

The engine then would move through the bay, starting its horrendous blaring and beginning to pull out of the driveway—while the rookie was still trying to get hands through his coat sleeves.

"Rookie! What the hell?" the captain would prod his new man, looking over him as he stumbled out onto the driveway with a shiny new helmet in his hands. The captain would then point to the fire station's address number—1500 West Wall Street—and the engine's siren would wane as the driver would run a hand round and round on the steering wheel to move the engine back toward the station. The raucous laughter would accompany the hooting hollering of the fellow firefighters, as loud as the engine's horn as it motored back into the bay.

The rookie—if he could take the bristling test—would smile a broad grin and shake it off. In all likelihood, he would be the one instigating the same initiation ritual with the next new man joining his crew.

Although the man they rib at breakfast may tow their water line at a fire that afternoon—watching their back as they hosed down snaking orange flames with hundreds of pounds of pressure shooting from the nozzle—knowing how much the man backing you up can take is as valuable as knowing how much one can handle oneself.

To the everyday working man, whose main goal might be dodging the boss at the office or setting daily goals on tasks while stuffed into a cubicle, the notion of such antics may seem, on the surface, to be useless harassment. To firefighters, however, the effort showed them how much each man could handle before their knees would buckle. If they could not handle the needling of a few pestering coworkers, they would certainly struggle with the force of a fire hose belching out water at three hundred pounds per square inch of pressure. Finely tuned corporations can work without such self-administered survival mechanisms, but the brotherhood of first responders needs a measure of its own to monitor the resolve of someone who must be trusted when life or death hangs in the balance.

"Doctor Jack" Williams was known at the fire department for his lack of tact. He fostered a level of sensitivity otherwise seen only in an easily angered West Texas rattlesnake. He vaguely resembled O'Donnell—primarily with his Texas twang and a wiry frame—but he lacked O'Donnell's height. Finding small talk a completely useless part of human interaction, Williams was disgusted with other firefighters on his shift at MFD Station 2. And they were disgusted with him. Each took to opposite ends of the station house until the alarm signaled a run.

As O'Donnell steered his brand-new Jeep Scrambler, polished to a glossy red, Williams looked on with several other firefighters at the beginning of a shift. O'Donnell smiled through a gritty black dip of Copenhagen as the early morning dew already began its burn-off. O'Donnell's uniform, as usual, was tucked neatly and tightly into the trim waistline of black, creased pants.

"You know, Robert, that's a real nice Jeep," Williams said, likely catching O'Donnell in a skidding surprise considering the comment's source. Despite detecting the coming sarcasm layered deep beneath Williams's compliment, perhaps, O'Donnell waved off concern with an eagerness to gloat over his newly acquired vehicle, purchased with proceeds from his new movie deal.

"Thanks, Jack," O'Donnell replied.

Williams cracked a smile through thin lips.

"And it fits the situation too," Williams said.

"Why is that?" O'Donnell asked, stopping to glance back at his new ride to examine just what he missed based on Williams's statement. "What do you mean?"

CHAPTER 20

Williams smiled and hedged slightly before continuing, "Well, it's a red Jeep," Williams answered. "Red—just like that blood money you used to buy it."

O'Donnell turned red himself, his mouth gaping, but fired up enough to get in Williams's face to contest the accusation.

O'Donnell's sudden "fortune" inflamed an already sensitive staff of firefighters who had to look on as O'Donnell's movie deal took flight. The deal had yet to pan out in those first months, when he showed up in a new car and boasted of his new home. But when producers finally did make the exchange of rights for money, only a handful of rescuers actually saw any cash. Although O'Donnell defended himself, saying the money he earned came from the fire department, few of the firefighters could imagine he had not sold interviews to news outlets—many did not want to believe otherwise and did little to rectify the inequity of rumor.

"To this day, I have not sold one article," O'Donnell told the *Odessa American* a year after the rescue, citing the rebuke of his fellow firefighters. "Everything I've done is for free. So these accusations people have spread are totally untrue. I haven't made a cent."

It did not help that O'Donnell's new Jeep had come with the purchase of the new brick home—in a more affluent part of Midland where the residual incomes of the oil field had otherwise dwindled, causing former upper-middle-class families to disappear. As the latest bust sank in, O'Donnell moved up and into Midland's expanding northwest development area.

The banter aimed at O'Donnell continued to build and gradually changed tone. His fellow firefighters judged the way he handled calls from the media as well as their own pestering, and the playful banter turned into spiteful chastising. And when they broke O'Donnell, he shattered, in several instances breaking into tears.

It did not help that O'Donnell felt threatened. On a late night in the days after the rescue and following his purchase of the new house in northwest Midland, O'Donnell, unable to sleep, fiddled with his motorcycle in his garage, taking it apart piece by piece before putting it back together. Hearing a thunderous crash in the driveway, he ran outside to see a large gash in the windshield of his Scrambler. It had been parked in the driveway facing the garage, not facing the street, O'Donnell looked at the brick that had been thrown through the windshield and considered the effort the culprit must have been willing to go through to do such a thing.

O'Donnell did not suffer alone.

Mine expert Lilly returned to Carlsbad with pats on the back and words of congratulations. It did not last.

A year after the rescue, Lilly sought a transfer. Though he shunned David Eyre when he came to town researching for the movie, coworkers accused him of violating ethics rules. Meeting President Ronald Reagan and appearing on Oprah's show fueled reproach among Lilly's coworkers.

Just a week after returning to Carlsbad, he was met with a malicious chiding. Coworkers alleged Lilly received promises of promotions and cash from private companies. He hired an attorney to sort it all out for him. Department of Labor standards kept Lilly from benefiting solely from his experience. He had worked at the scene as a government employee—and so it was that Uncle Sam owned half of the experience.

If he knew what Midland would bring on October 15 as rescuers clamored for a mining pro, he would not have made the 20-minute ride on Price's overpowered jet across the desert. His concession, though, always went back to picturing Jessica and his hand in her rescue.

It was worth it, he would conclude.

The trouble with coworkers continued for O'Donnell. While Lilly eventually nailed down a promotion and moved to Alaska—after initially getting turned down for a promotion he thought he deserved—O'Donnell's options were much more limited.

Coworkers found every opportunity they could to rub O'Donnell's nose in his screw-ups. They got plenty of chances.

CHAPTER 20

When the city's Department of Health officials learned of the state's effort to treat country rodents for bubonic plague by dropping boxes of pellets from the sky, they called fire stations, warning them to be on the lookout for bubonic plague symptoms. O'Donnell took the call at MFD Station 2.

>Sudden onset of high fever.
>Chills.
>Muscle pains.
>Swollen lymph glands.

To notify fellow paramedics, O'Donnell scribbled out the symptoms on the chalkboard. BE ON LOOKOUT FOR BLUEBONNET PLAGUE. The alert for bubonic plague turned into a confused warning for some strange plague named after the state flower of Texas. "Doctor Jack" Williams and several other firefighters walked in to see O'Donnell's message. Dumbfounded either by O'Donnell's complete lack of medical knowledge or what they thought might be a ridiculous joke, Williams chalked up O'Donnell's note to a heightened sense of self-importance and likely a symptom of his now well-known consumption pattern.

Williams edited the alert with an additional note to accompany O'Donnell's "Bluebonnet Plague" message: IT IS BROUGHT ON BY THE PITIFUL ORGANISM 'ROBOTICUS HEROTICUS' AND CAUSES A LAPSE IN UNDERSTANDING OF HISTORY AND MEDICINE. The crew broke into laughter as O'Donnell sat awash in his coworkers' chastising censure.

CHAPTER 21

AS FAST AS HE ROSE, HE FELL

ROBBIE O'DONNELL'S MOTHER CALLED HER AT WORK ON Monday morning, April 24, 1995.

"Well, he finally did it," Joanne Martin said on the other end of the line. "Robert finally did it."

"Did what?" Robbie replied, expecting some strange anecdote about her former husband.

"He finally killed himself . . . last night," Joanne said.

Robbie could only think of one thing.

"What am I going to tell the boys?"

She drove all the way from her office in downtown Midland to Greenwood—about twelve miles—where fourteen-year-old Casey and ten-year-old Chance attended school.

She stopped to talk to Casey at the junior high school first.

"Baby," she said, her entire life seemingly leading to this fateful moment, defined by actions of others but bearing a heavy burden on her life and the lives of the children she help make. "Dad killed himself last night."

Casey huffed, bordering on a sigh.

"Man, what are we gonna tell Chance?" he groaned, having taken on so many duties already as the man of the household.

Still, the question was a good one and one for which Robbie had not a clue as to how to answer. Chance looked at life a little more like his dad did. Though Chance did not favor his dad physically in appearance, he was his father's shadow, picking up dirt bikes at a young age and taking on a number of his traits. For a few months while his dad lived in Huntsville, Chance lived with his father after considerable pleading and begging. Chance often called Robbie, crying about his father. Working long hours, Robert was not there to cook dinners. He was rarely sober.

For a son, wanting to live with his dad was natural. Robbie knew it was a grass-is-greener scenario and relented.

"He had to see what it was like for himself," said Robbie, revealing why she allowed her son to live so far away with his prescription-drug-abusing father.

After those few months, Chance called Robbie, ready to come back to Midland.

And on that Monday morning, while students tinkered in classrooms nearby, Robbie and Casey had to work out a way to tell Chance that his father killed himself.

"You better let me tell him," Casey insisted.

Chance popped out of class and into the hallway in high spirits and wondering why he was being freed from school so early. Robbie and Casey put off telling him anything, hoping to make it to her mother's house before they had no choice but to tell him.

Pulling into the home's driveway, Robbie shifted the car into park and Casey looked at her with dead eyes.

"Mom, just go," he said, motioning for her to leave. The nature of their world suspended, two brothers had to live this moment together, mustering the strength from somewhere deep within to bear the brunt of their father's death at his own hand.

They all got out of the vehicle. Robbie stood in the garage shaking and crying, her whole body starting to become sore from all the worry and anguish. On the side of the house in the middle of the day—a day that would have been like any other in Midland—Casey told his brother that their father had killed himself. Robbie turned the corner from the garage to see her sons in a hard embrace.

The days passed by between getting the body back from the medical examiner and preparing for the funeral. Robbie and the boys learned little of the actual circumstances behind the suicide. They did not need the details.

In the aftermath, a whirlwind of accusations swirled around the Midland Fire Department—especially around Chief Roberts.

Roberts issued assurances that he had done all he could for O'Donnell. Many disagree today, recognizing O'Donnell's cries for help—help in understanding what was wrong, help in kicking his addiction to prescription meds, help in getting his life together. Maybe he did not know what he needed. Most today still do not know what he needed, but they think someone should have helped him find out.

Chief Roberts, according to a number of sources, did not attend the funeral.

"The Dance" by Garth Brooks echoing its heart-wrenching effects rendered the funeral a jumbled mess of emotions.

Cissy McClure attended, as did Midland County Sheriff Gary Painter. And so did the media.

Some waited outside the gates. Others went in. One female photographer approached Robbie as she and her boys walked to their vehicle to leave. Robbie, feeling the photographer right in her face, smarted off.

"And there's part of what killed him right there," Robbie said.

The photographer fired back. "Excuse me? I'm just doing my job here."

Robbie surged, "Look, lady. Get that thing outta my face."

"You better listen to her, lady," one of Robbie's boys encouraged.

The photographer edged closer, prompting Robbie to rare back with a fist and slug the woman. Recoiling, the photographer saw Glasscock in uniform and walking behind Robbie's group.

"I want to report an assault," she called out.

Glasscock looked at the photographer, shook his head, and said, "I think you better leave."

Robbie later recalled, "Man, that felt good."

CHAPTER 21

At 2 a.m. several weeks after the suicide, Robbie awoke to the pinging bounce of a basketball on the driveway. Trying to wake up and determine whether she was actually dreaming, she looked out her kitchen window.

Casey stood in the shadows, bouncing the basketball, shooting baskets—and bawling. His sobs were covered by the constant bounce and clinging bang of the basketball on the hoop's rim.

In dealing with the death of their father, Robbie did not want the boys to bottle up any more emotion. At night before bed, she would call the boys into her room, piling them onto the bed.

"Okay, what are you thinking today?" she asked over the complaints and groans of adolescent boys being called on to share their emotions.

"Nothing," one would say.

"I don't even care," the other would tersely reply.

One of them would eventually crumble, giving up some part of their emotion. Then the other two would fall with them. They sat on the bed, banging their fists into the pillows, yelling into the fluffy mass to absorb their anger.

The last thing Robbie wanted was for them not to express their anger—not to suppress their emotion. If so, it would mean repeating the mistake allowed of her ex-husband.

"It's okay," she would tell them. "Be mad."

There are aspects of the aftermath that continue to anger Robbie, though. While most of her feelings toward O'Donnell are rooted in the fact that he is her kids' father, she looks at him as a hapless wanderer.

Still, there are those who could have helped, she maintains.

Living in a small town with small-town gossip and small-town circles did not help matters. It meant constantly seeing people who reminded you of the things you were trying to avoid. It meant road markers and benchmarks pocked your mind like land mines waiting to go off. A simple song on the radio or a billboard could ignite a far-off memory. And sometimes, a small town just meant you saw people more regularly than you would like to, bringing you face to face with those you know bear at least a little responsibility for your plight.

AS FAST AS HE ROSE, HE FELL

While working for Wagner and Brown, a local Midland oil company where she looked over oil deals and land agreements all day, Robbie took a break in a shared break room where employees from all over the building gathered. Chief Roberts, who was overseeing the architectural schematics on a new fire station in the planning stages at another office in the same building, stopped in the break room for a cup of coffee. The two locked in an awkward stare before exchanging pleasantries.

Robbie turned to leave.

Behind her, she heard the chief.

"Robbie, do you think I am responsible for what happened to Robert?" he asked.

Robbie, always polite and gentle in person—except apparently when being jostled by a wayward photographer—opted out of the confrontation.

"Did you pull the trigger?" she asked and walked away.

She tells the story now, rolling her eyes.

Today, her neighborhood is quiet. Within sight from her front yard, where she spends spring days in the garden, is the home where Jessica McClure grew up. It is just a few houses down and across the street. Robbie now works for another major oil company in downtown Midland.

Occasionally, Cissy drops by with a pot of potato soup or another home-cooked goodie. Robbie and Cissy sit at the kitchen table and chat—not about anything in particular, just a chat like two neighbors whose lives never intersected over anything more than the proximity of their residences. They each received a call at least once from a producer with the *Dr. Phil* program, hoping the women would consent to a show on their lives.

This life, with calls coming from all over the nation—often on anniversaries marking one significant event or another based on the rescue—will always include persistent pestering.

Robbie has learned to deal with it.

"It'll never end," she said. "It will never just be over."

CHAPTER 22

"I DON'T KNOW WHY I DID WHAT I DID"

"WE WERE JUST A COUPLE OF GOOD OLD WEST TEXAS boys who did their jobs. In the end, we never saw what was coming." Andy Glasscock puts almost twenty years of turmoil into a sound bite, spilling out the words to a reporter who still wants to listen.

His speech gushed out like it had back on June 29, 2004, when he stood before US District Judge Robert Junell, sniveling his apologies before accepting a sentence of fifteen years in prison for his foray into the world of child pornography.

"I got off into a sexual addiction like alcohol and drugs," Glasscock blubbered. "I never meant to hurt anybody. I would never hurt a child. In my heart and soul, there is still good left in me."

Judge Junell, sitting at the bench and looking on as Glasscock continued, replied, "Absolutely."

"I have no excuse for what I did. I'm deeply sorry. I never meant to hurt anybody. I'm not a sexual monster. I don't know why I did what I did."

In May 2005, less than a year into his sentence, Glasscock was ready to criticize the efficacy of the federal prison system's rehabilitation program. He maintained he would have had a better chance of improving mentally and becoming rehabilitated had he been allowed to participate in a community-based counseling program. He remained baffled by the prosecutors' lack of consideration in their sentencing recommendations for his community service record as a peace officer.

"To sit in a courtroom and have this young kid federal prosecutor trash your life—yeah I had some problems, but I'll tell you what, there's a

CHAPTER 22

lot more people that have those same problems," Glasscock said. "Maybe that's my calling. I could have walked out of that courtroom that day and got some psychologist help and never done it again."

Glasscock had lost everything. Nothing monetary remained his own.

From his federal prison cell, he looked back on the days of his police career, and he was disappointed by his power-hungry fits and mad pursuits of adrenaline. He thought back to all the times he was given a second chance and how he screwed those up, too. And he thought about how he could have done so many things differently. He thought of those follow-up interviews from a screeching press pool, those calls he wished he had ignored.

The same calls went to Forbes, who took the handoff from O'Donnell at the bottom of the well, and whose claim to the rescue's fame is just as strong. Forbes today is captain at Station 3 in the MFD.

Back when the rescue of Baby Jessica was wrapped up in a banner-headline victory, all the rescuers returned home—Glasscock, Forbes, and O'Donnell.

O'Donnell found his home late that Friday night full with Robbie's family and his excited children who could make little sense of what their father accomplished. Robbie's mother clicked away on a camera, capturing O'Donnell, his shirt half-buttoned, a mesh cap still crooked on his sweaty brow.

The phone was already ringing.

Even as Robbie ran out into the front yard to hug her husband, she told him of a call waiting on the line for him.

"It's some radio station in New York," she told him.

"What? Are you kidding me?" he joshed back, still carrying a bottle of Dr. Pepper.

Lifting the receiver from a countertop, he answered, "This is Robert." His drawl, spanning close to a thousand miles in an instant, reached out to a world that moments before had no clue who he was.

The voice on the other end was a fast-paced, excited, and congratulatory caller from a city O'Donnell would likely never visit. Still, the caller,

"I DON'T KNOW WHY I DID WHAT I DID"

as personable as someone O'Donnell had known for years, wanted to hear how it felt to save Baby Jessica: Everybody's Baby. The caller wanted to know for himself and for thousands of his listeners tuning in all over New York City.

"Yeah, sure, I'll talk to ya," O'Donnell said.

When he hung up, the phone rang again. And again. And again.

EPILOGUE

Robert O'Donnell

The green glow of the instrument panel lit O'Donnell's weary face. In the dark, vast desert near Big Spring, Texas, he reached for whatever scraps of paper he could find in the cab of the pickup, pressing them on the steering wheel and looking for a pen. The one he could find dribbled black ink, spilling all over the back of a crinkled receipt for incidentals at Koko Inn, where he had stayed in Lubbock while attempting a start in a new job and a new town. "Mom, give this to Casey and Chance."

The words are mangled on the slip, much of the writing illegible, making it difficult to tell whether the marks are the result of exhaustive emotion or prescription medication. "I am very proud of those two."

The last readable word in the note before his name at the bottom is "future." Little else can be made of the note though there are other trickling lines of ink.

In his brand-new pickup, O'Donnell reached across the seat for the shotgun.

William Andrew Glasscock

It would take about twenty minutes for Andy Glasscock to finally come see you after you summoned him from his cell. The guards knew you were coming, but they took their sweet time to call him to the visitors lobby. It is not like he was going anywhere, anytime soon.

Serving time in a federal prison on the East Coast, sentenced June 29, 2004, on charges of sexual exploitation of a child and improper storage of explosives, Glasscock worked for nine cents per hour handling the recycling at the prison.

EPILOGUE

His family never visited, he said. The trip was too far. The federal prison is a several-hours-long drive from the nearest major airport. The only time they had seen him since being found guilty was in a few photographs he sent from what appears to be a small corner in a prison yard. On Tuesday, July 12, 2005, when he pleaded guilty to state charges of raping Sandy Lou Smith, he received a concurrent term of twenty years in prison. He received credit for the time he had already served in federal custody.

When it came time for an allowed victim impact statement, Sandy Lou Smith, a pseudonym assigned by law enforcement, had taken the stand.

"Could you speak up, please," he requested in the court as she started to speak.

She replied tersely. "You became a police officer so you could commit these evil acts. Just the fact that you committed this heinous act tells me you're the kind of person who doesn't care."

His appearance in 2005 was much different from the year before, when he blubbered his way through an apology to US District Judge Robert Junell. His hair, then silver, was combed from the left side of his head in a large wave, matching his long gray beard.

His federal term and state term ran concurrently, which at the time seemed to offer the possibility of one day seeing freedom once again.

As of 2005, no one in prison knew of Glasscock's past. No one knew he had served as a police officer who could be like one of those investigators who had put any one of these inmates behind bars. Maintaining cover was not so simple.

"As you can imagine, we watch a lot of BET [Black Entertainment Television]," Glasscock said, seeming to make a joke, though he is quite sullen in a khaki uniform barren of any real color.

He explained how during a documentary of Michael Jackson, inmates gathered around a television. The King of Pop's catalog of music videos flashed on the screen intermittently. Suddenly, Andy Glasscock is on the screen, appearing for an instant in the video for "Man in the Mirror," which features the shot of Steve Forbes sprouting from the earth with Jessica McClure in his arms.

EPILOGUE

He froze, his eyes darting around the room, checking to see whether anyone somehow recognized him in the split second he said he saw himself on the screen.

Hours after sentencing by Judge Junell, Glasscock was sent back to an Odessa jail to await his transfer into the hands of the US Bureau of Prisons. Though Odessa protected Glasscock, letting him use a pseudonym, an inmate also heading off to federal custody approached the former cop. "Hey, Glasscock, what's up?" He froze with fear that his time in safe custody might be limited.

In prison, life expectancy for a police officer is short. Expecting to be placed at a federal prison in Butner, North Carolina, for its renowned sex offender treatment programs, Bureau of Prisons officials ignored his pseudonym as they carted him from Odessa to Big Spring to Lubbock to Oklahoma City.

"Glasscock!" each guard would shout to summon his presence, negating any chance of securing his anonymity.

Staying a night in Lubbock, Glasscock was placed in the Lubbock County Jail's already overcrowded general population before he could get the attention of one of the female guards.

"I used to be a cop," Glasscock tried to explain, hoping to be placed in a private cell or any place that might protect him if he were recognized.

At the time, Glasscock thought he was on his way to Butner. He did not know, however, that the facility was full and that he would be headed elsewhere. The utility of his status as a law enforcement official was long gone, and the privileges, no matter how they had served him before, had become as insignificant as he felt about his place in the world.

"The Butner institution has an extremely high population and is presently unable to accommodate additional inmates," Ronald G. Thompson, regional director of the Bureau of Prisons, wrote in a letter to Junell, dated July 26, 2004, prior to Glasscock's state sentencing. "His assignment to the Memphis institution places Mr. Glasscock in a facility, which meets his security needs."

Glasscock kept to himself for the few hours he was in general population in Memphis. Word circulated quickly among inmates of the former cop in their presence, thanks to the quick public-relations work of the Odessa inmate who recognized Glasscock.

On the telephone with his son, his eyes darting all around him just waiting to get shanked, Glasscock suddenly felt a pointed nudge on his shoulder.

"You need to get off the phone now," said an assertive guard who heard the circulating word of a price on Glasscock's head. Glasscock spent the rest of his time in Memphis in solitary confinement for his own protection from the general inmate population.

He eventually made his way east, though his concern for his anonymity in prison brought him to beg for a limited release of information regarding the location of his imprisonment. The same specifications existed for his appearance.

Somehow, Glasscock's ability to remain "just another inmate" allowed him to live. His experience is one that has opened his closed eyes, he said, perhaps shedding light on a dark place where he once lived as a free man—as a police officer. And so, it is then that what saved his life is what almost ended it—prison. He wished that he could have been a different kind of police officer. He wished he had listened more to suspects, been less invasive, been more understanding, been less like a cop who sees the world in black and white.

He said a number of fellow inmates were better people than the police officers he worked with in Midland. But now, words were not those of a respectable, upstanding citizen. They were the words of a criminal. And people back at the MPD would slough them off that way.

On July 10, 2016, Glasscock died in prison. His funeral was held in Midland five days later, and he was buried some sixty miles away in Big Lake's Glen Rest Cemetery. He had eleven grandchildren and step-grandchildren.

Scott Shaw, Photographer at the (Cleveland) Plain Dealer

He looked grudgingly at a couple bottles of champagne chilling on a table near a spread of party snacks, cakes, and cookies.

EPILOGUE

"This is going to stink if I lose," Shaw thought to himself.

It was a key moment in Shaw's professional career, although the significance of the day does not fully commit justice to the reality of the looming prospect for his future success as a photographer. A call away was the Pulitzer Prize for the photo he snapped of Jessica McClure as she emerged in the arms of Steve Forbes. The call could also be bad news, though, and as Shaw looked at the chilling bubbles, he pictured having to soak his misery in the frosty drink.

He was up against two other photographers, both of them previous Pulitzer winners.

"I figured one of them would win another one," Shaw said. "I didn't really get my hopes up."

Among twenty pictures submitted by the *Odessa American*, Shaw's shot of a number of rescuers looking right at McClure as she emerged was the only one in which his subjects looked as though they were in perfect coordination.

"It was like I told them all at once to turn and look at Jessica," said Shaw.

With all the photos submitted, however, Shaw thought he had less than a remote chance of getting his photo selected. It was overkill, he thought.

So, when a Pulitzer committee selected the metro section photo as a finalist, Shaw certainly was surprised. Over in Odessa's sister city, the *Midland Reporter-Telegram*'s own work fell short on the prize-seeking.

The competition between the warring neighbor cities fell again to Odessa—the gritty town of roughnecks compared to the slick-living Midland snobs. The *Reporter-Telegram*'s editor sucked it up, penning words of congratulations to *Odessa American*'s editor Olaf Frandsen before the Pulitzer photos would even be submitted.

Nov. 4, 1987

Dear Olaf:

Things have been unusually hectic here in the days following the rescue of Jessica McClure. Hence, the delay in letting you know how impressed I, and all members of our news department, were with Scott Shaw's outstanding photos of the rescue.

Both your Page 1 photo and the large Page 3 photo of October 17 are prize winners and, as I told my staff as soon as I saw them, "They will win whatever they're entered in, possibly including the Pulitzer."

I understand the latter is among areas in which this photography will be submitted, and we wish you and Scott well.

His work in this coverage deserves any awards it garners.

Sincerely,

Jim Servatius

A phone rang in the newsroom, prompting a reporter to pick it up as Shaw stared down at the frosty bottles of bubbling champagne. The newspaper's publisher continued hustling back and forth to check his computer for updates, though no word came across the wire, and word on the prize was set to be announced any minute.

Local television crews waited in the newsroom with beaming lights settling beads of sweat on Shaw's head. Supporters continued pestering him with expressions of their confidence in his photo.

The reporter fielding the call talked for a second.

"Who is it?" someone asked.

"It's UPI radio wanting to talk to Scott," came the reply.

"Why do they want to talk to him?" someone else asked.

"Because he just won the Pulitzer Prize!"

The crowd swarmed Shaw, splashing a shower of cold champagne all over him.

Midland Police Chief Richard Czech

The police chief who came to Midland from Arizona with an eye on reforming the police force made a large mark on the department during his stint, which lasted years. Police say Czech lobbied hard for John Urby to replace him as chief. Czech left the department better than he found it, the rescue of Jessica McClure little more than a footnote to his work in Midland.

EPILOGUE

Though he was depicted in the movie *Everybody's Baby*, it is not known whether he accepted what would have been a $7,000 payment for his likeness.

His gruff demeanor and no-nonsense attitude came across quickly in March 2005 when he was interviewed for this book. Maneuvering around pointed questions, he always returned to the public-relations glad-handing that shrouded the many effects of the rescue. He maintained composure and remained honest about facts. Still, he did not seem to grasp the psychological repercussions that plague men in his profession.

Midland Fire Chief James Roberts

More than just the family of Robert O'Donnell carry a disdain for Chief Roberts. At least one other firefighter injured on the job claims that Roberts turned his back on him in his hour of need, after he was badly burned in a fire the MFD understaffed. That firefighter claims O'Donnell suffered the same kind of disregard from Roberts and the department's leadership.

Roberts's wife screened all calls to their home.

"The story has been done enough. Everything that can be said has been said," she claimed.

That response did not surprise the firefighters and O'Donnell's family, who claim Roberts kept his cards close to his chest and sat "too tight" with city council. Though that ability to play politics helped yield the fire department equipment to stay ahead of the game, Roberts's critics said it required a costly tradeoff—one that kept him from getting close with his own firefighter staff.

Roberts retired in the late 1990s. His home is tucked in a quiet neighborhood in central Midland.

In a follow-up story covering the ten-year anniversary of Jessica's rescue on *NBC Nightly News*, Roberts said the salvation of the young girl capped off the story, seeming to also make a statement about the stories that came after.

"She's walking around the streets of Midland today. And you can't take that away from us. We did it," he said.

Paula Bynum

Paula Bynum was fired from the MPD soon after an official investigation of her self-inflicted gunshot wound. Though reports are conflicting, some in Midland say they have seen her working at PetSmart and others say she's been seen working at Walmart.

According to county property records, Bynum still lives in the same house she lived in when she was fired by the Midland Police Department for the shooting incident.

Bill Bentley

Bill Bentley, the caver and Dimension Cable employee who stayed at the scene and who donated his caving equipment to assist in the rescue, now works for Cox Communications. His vast collection of Jessica McClure rescue memorabilia includes a large number of videos and articles.

He operates a caving website with links to a page dedicated to the Jessica McClure rescue that includes photos and links to articles.

Dave Felice

Dave Felice, O'Donnell's friend and captain, now lives in State College, Pennsylvania, where he works as a fire inspector.

Felice left the fire department in the early 1990s.

He last spoke with O'Donnell just a month before O'Donnell killed himself.

Patrick Crimmins

Crimmins left Midland in the early 1990s for the *San Antonio Light*, the now defunct paper absorbed by the Hearst Corporation's *San Antonio Express-News*.

Crimmins eventually went to work in public relations, taking positions with several state government agencies before starting his own PR organization in Austin.

Steve Forbes

Steve Forbes quickly began opting out of interviews, starting less than a year after the rescue. In the *Odessa American*'s one-year anniversary spread on the rescue, an in-depth article on O'Donnell noted Forbes' having turned down the request for an interview.

O'Donnell, on the other hand, landed a 30-inch story in the section.

Forbes became a captain at the Midland Fire Department.

He retired as a firefighter in January 2014. The retirement was heavily covered by local news media.

Cissy McClure

Cissy McClure today owns and operates a pet grooming shop across the street from the old Dellwood Mall in Midland. Friends say she remarried at the roadside park where she met her current husband in the early 1990s.

When approached for an interview about this book, Cissy replied, "No. No uh-uh. I don't I don't. No."

She and Chip divorced in 1990, and she remains in the Greenwood home she is rumored to have been afforded by the thousands of donations sent from all over the world.

The O'Donnell Family

After her ex-husband's death, Robbie concentrated on her sons, Casey and Chance. At the time of Robert's death, Casey was ten years old, and Casey was fourteen years old.

In the early 2000s, Casey O'Donnell followed a version of his father's path into emergency response. As a sergeant deputy in the Midland County Sheriff's Office, Casey routinely patrols Midland County, responding to criminal activity and answering emergency calls.

In 2017, he appeared in an episode of A&E's *Live PD*, a reality show depicting live action on the streets with officers responding in real time.

Chip McClure

Chip McClure went into real estate after leaving Midland. After several marriages and a move to Tyler, Texas, he started a career brokering jet aircraft. He has since remarried happily and lives in Tennessee. In 1996, he participated in the development of a book, *Halo above the City*, about the rescue. He helped author the book with Tyler real estate agent Teresa Farish. In 2023, he said in an interview that he had not intended for the story to be told in the way that it came out, attributing many of its flaws to a well-intentioned but inexperienced writer. He was disappointed with the outcome.

Rumors dogged Chip for years. One included his confinement in the local Midland County Jail on the morning the rescue started, accounting for the delay in getting him to the scene of the rescue. Other rumors became so pronounced that even his close friends came to imagine a new reality for Chip. In one instance, Chip explained, former Midland County Sheriff's Deputy Mike Hall had become a close friend who would come by a trailer manufacturer shop location where Chip eventually worked. Hall would chat casually in a group of Chip's coworkers. "He said, 'Man, you really have come a long way, and I can't be more proud of you. I've seen you at your lowest and now, just man, you've got your life together.'" Chip recalls the moment with a profound shock washing over him. "I looked at him, and I said, 'Mike? See me at my lowest? What are you talking about?'"

"I think it was sitting on your butt in the county jail," Hall explained.

"Mike, I have never been arrested. I have never been in your damn jail. I've never had anything more serious than a speeding ticket," Chip said, perplexed. "No wonder people believe that damn rumor if you're going around as the deputy sheriff... if you're going around telling it."

In the years that followed, Chip began detailing his account of the rescue on Facebook. The posts generated row after row of replies from followers. Many of the commenters recalled their own experiences watching the rescue unfold on TV, vividly outlining their memories of the harrowing ordeal.

Jessica McClure

Jessica McClure graduated from Greenwood High School in 2004. In 2005, she was living in an apartment in Midland and reportedly attending Midland College. The only public statement she had made—besides

EPILOGUE

at least two she and her mother granted in exchange for money to *Ladies' Home Journal*—was in a letter to the editor to the *Odessa American* in response to the continued development of a nuclear waste storage facility in Andrews, Texas.

At 1:30 p.m. on a hot day in June 2005, she answered her home telephone, groggy, just awoken from a nap. On subsequent attempts to reach her by phone, she would not answer. It was the only time I got her to respond to any questions.

The glow from the scene and the shock of those round eyes peering into the night may be the best way to remember her story. In the years since, she has gone on to hold multiple interviews on television and in print. She admits having no memory of the rescue. She lives in Midland with her husband, Daniel Morales, and several children. In 2011, reporters covered the benchmark that was her twenty-fifth birthday, effectively opening access to a trust reportedly opened for her to handle the funds sent to her from around the world. Midlanders speculated for decades on the exact total, rumored to be in the millions. In 2017, *People* magazine interviewed Jessica. Writer Darla Higgins reported on the trust's tally as well as on Jessica's perspective on the rescue.

> While some have assumed that Jessica became wealthy as an adult, thanks to a trust fund set up by people from all over who watched the rescue unfold, she says much of the $1.2 million in the trust disappeared during a stock market dive in 2008. Much of the remaining funds went toward buying the family's modest house, which features a large backyard for the kids.

In a one-year anniversary profile of the family by *People* magazine, the McClures confirmed their cash purchase of a $30,000 home outside Midland and a new "custom pickup and a shiny black Thunderbird." They went on to discuss the trust established for their daughter but not with enough specificity to confirm the rumored $700,000 balance. Rather, they acknowledged a financial security Jessica would enjoy. "We will only use [the donations] for medical expenses directly related to

Jessica's injuries or for her education," Cissy told *People*. "She'll go to a private school. I want her to get a good start on her education. I didn't get a good start on mine."

Midland commemorated the rescue with a bronze plaque tacked to a wall outside the Midland Center where Oprah Winfrey filmed her show to honor the rescuers. Constructed by Midland sculptor Mary R. Griffith, the piece was underwritten and donated by Mr. and Mrs. Jim Cocke of South Padre Island. The two had read about the Jessica rescue after it unfolded while they vacationed on safari in Africa.

Inscribed on the plaque are the words: Nothing the heart gives away is gone. It is kept in the hearts of others.

Griffith aimed to capture in bronze only a few of the faces of those actually at the scene. Jessica McClure. Steve Forbes. And, though he remained at the bottom of the rescue shaft as Baby Jessica was lifted amid grateful shouts, Robert O'Donnell.

ACKNOWLEDGMENTS

I AM SO GRATEFUL FOR MY WIFE, MICHELLE. SHE endures my desires for new challenges with interest and encouragement. I should also make a special note of appreciation for my parents, who more than anyone always believed in me even when I had fallen. Their work proved to me early that it was important to get back up after you have been taken to the mat.

It's easy to look back and see how your experiences shaped your thinking as you progress into adulthood. It's another thing, however, for me to look back on the documented work I produced as a young reporter. I knew that a variety of different perspectives would come along with twenty-plus years of fatherhood, marriage, and career development, but it was interesting to return to these pages and notes with new angles that came with time.

I hate admitting it.

I have to concede the importance of the passage of time as you invest in knowledge and learning from others around you.

Early in the research around Robert O'Donnell's death that formed the incipient stages of this book, one of the sources highlighted just how the advantages of experience and age reveal themselves slowly over time. He described the moment when Robert O'Donnell reached Jessica McClure in the well. In that moment, O'Donnell could not quite extract the child. Her touch near the rescue's end sent O'Donnell into an emotional tailspin, my source said. At the time of this interview, I was a young new father, and suddenly, a wealth of emotions swept over me as he detailed what O'Donnell dealt with. Inexplicably and without warning, I felt the pain O'Donnell must have felt as he stretched and reached for Jessica. It took me a minute, but as I felt my throat tighten and my eyes well up, I realized what was happening. Synapses fired, setting off a chain reaction from a mental picture of that instant. It connected me to O'Donnell as he struggled. I could suddenly feel what he felt and hear what he heard. The soft touch of her skin. The whimpers in the darkness. In envisioning that encounter I had imagined my own daughter. Being a young father responsible for a human life that depended on me with every living breath had given me a perspective that would have otherwise simply been that of a set of facts from a source.

Certainly, experience had affected my perspective.

As a young reporter and writer, I remember a hunger and thirst for success that was often overwhelming. At the time, those editors and mentors in my circle tried to settle me in for a long haul with assurances that time would reveal opportunities for success. For some of those extraordinarily talented writers and reporters, a long timeline of accumulated work might not be necessary for insight, but looking back, I can see how it gave me additional perspective. I can only imagine what another twenty years might provide.

I was lucky to curate a circle of friends and colleagues who allowed me often to indulge my curiosities and act as an interested audience as I developed a sort of factory for pursuing these interests—from stories to businesses to new journalism technologies. They often entertained ideas and projects with an active presence, but it would be foolish to ignore how often they had to look the other way when my foundry faltered embarrassingly. Those with a shared curiosity never seemed to question me when I returned to the factory floor to get the production back up and running. I am grateful for those kinds of thoughtful friends.

Among them are fellow reporters Cory Chandler and Eric Finley. Their talent often went beyond my own, but they were always trustworthy sounding boards interested in what antics I employed. Good humor with an appropriate degree of self-deprecation goes a long way toward making the never-ending days in the early stages of a career go by

ACKNOWLEDGMENTS

more quickly. Other friends have taken me into their employ as I set out to try new career paths and take what I learned as a reporter into their businesses. They gave me new perspectives, shared their knowledge, and for the most part gave me the running room to test, re-test, and try new things. It took a great deal of trust, and I appreciate their willingness to let me swing for the fences.

I met Stephanie Limb following the 2013 Texas legislative session as a partner to take on healthcare policy communications. Much has emerged from what has become a strong alliance and business partnership. But as a friend, I gained much more. I have learned the value of an opposing, well-informed political view over one in agreement without regard for information, facts, or real people living complex lives with complex challenges to overcome. Stephanie is a willing ear with a cogent and agile mind that for some reason uses its Ivy League foundation to hear out whatever new story, idea, or question might be posed by this state college grad.

In 2005, I set out to put together the original pieces of this story with a focus on public service and those who truly set aside self-interests to help others. The scope immediately widened beyond that as I realized the extent to which some of those I began profiling had lost track of their original mission. I have met many political, corporate, and healthcare leaders over the years who keep their community service at the front of their minds. Many of them foster a set of skills and tools that proved to be critical in my career development and that I try to build in a way that serves others and improves my community. The best among these leaders, using data and personal narratives from their teams, can weave a story together to motivate and tap into shared missions, moving people forward to accomplish very difficult tasks and achieve goals. They ignore the desire to castigate their critics. They avoid ceding their limited share of voice to detractors and those who do not have a shared mission. I have appreciated these leaders, because the most successful among them realize that people are moved by stories, and when they properly prepare their audiences, they can unmoor them from the kind of cognitive dissonance that chains our voters, our governments, and our communities to the status quo. For those who see themselves as public servants, focusing on a true mission and sharing it with their teams can deliver what we expect of those who wish to truly serve communities. To have their voices rattling around in my head, I am grateful.

REFERENCES

ANY QUOTE OR CONVERSATION IN THIS BOOK HAS BEEN taken from interviews with those involved in the story as it happened from their own perspectives and recollections. Trustworthy dialogue and quotes are published as they were presented to me or as they were published in other source material such as newspapers, radio, television, and judicial records or proceedings.

I drew on a vast collection of newspaper articles from publications around the world, spending hours sifting through microfilm in the Midland County Library and the George & Helen Mahon Public Library near the *Lubbock Avalanche-Journal* newspaper offices in Lubbock, Texas.

The media, as depicted in this story as I chose to tell it, are cast when the opportunity presents itself as characters due to their relationship with the emerging twenty-four-hour news cycle. In many cases, I extracted dialogue from articles to create a voice from those around the world who were impacted by witnessing the event in real time. Any quote used in the book occurred as it appeared in various publications. Other cases include dialogue captured from interviews with television media, which is broken down and presented as conversation between reporters as characters.

Much of the material is sourced from court documents provided to me by individuals involved in the story or obtained through public information requests submitted to various government entities.

INDEX

ABC News, 158, 159, 184
Abilene National Bank, 30
Abra Kadabra, 212
ADMAC, Inc., 146
Alfred P. Murrah Federal Building, 8–9, 223
Allday, Martin, 207
Applebome, Peter, 197
Arrington, David, 28
Associated Press, 134–35, 197, 204

"Baby Jessica." *See* McClure, "Baby Jessica"
backhoe tractor, 38, 58. *See also* rathole rig; Green Dragon
Baker, Shala, 106–7
banking, and oil boom and bust, 25–27, 29–30
Bardach, Sheldon G., 200
Barker, Lisa, 160–61
Barker, Mike, 177
"beast, feeding the," 164
Belkin, Lisa, xvi, 204
Beltran, Manuel "Manny," 71, 72, 81, 82, 100
Bentley, Bill, 39, 59, 61, 84–85, 87, 90, 93, 248
Bible Belt, 131
"Big Empty, The," xi–xii
Big Lake, Texas, 23
Boler, Charles, 91, 190–91, 198
Boler, Floyd, 91
Boler, Larry, 91
Boler, Ribble, 91
Boston Globe, 148
Boyd, Sharon, 212
Bracken, Steven, 111, 118
breathing system, 73
Breckenridge, Texas, 23
Brooks, Susan E., 201–2
Brown, Rick, 19–20, 52, 207
bubonic plague, 230
Buksbaum, David, 182
Burnett, Warren, 11
Burns, Morris, 24–25, 26
Bush, Barbara, 215–16
Bush, George, 214, 215–16
Bynum, Paula: behavioral health of, 104–7, 122; discharges firearm, 107–8; and Glasscock divorce, 167; MPD coverup regarding, 108–10; post-rescue life of, 248;

and rescue of Baby Jessica, 68, 104; self-shooting of, 110–29; speaks with *Midland Reporter-Telegram*, 126–28

cable news, xvii, 151–53. *See also* CNN
Calder, Iain, 201
caliche rock, 60–61, 82, 91, 93
Capell, Larry, 45, 46–47
Carol, Karen, 110–11, 115, 123, 124, 129
Carter, Ken, 45, 47
Cartwright, Gary, 94
Cassidy, Dave, 98–99, 100, 135
caving, 39
CBS News, 154
Chambers, Stan, 165
Chandler, Dorothy M., 136–37
Chandler, Floyd, 136
Chastain, Gary, 47, 48, 49
Cherry, Chris, 110, 119–20
child pornography: and arrest of Andy Glasscock, 170–71; and sentencing of Andy Glasscock, 237
church, 132–34
Clark, Tony, 161–62, 176–77, 179, 180, 184
Clifford, Patricia, 207
CNN: coverage of Baby Jessica story, 161–62, 176–77, 178, 180, 184, 185, 186; news-reporting innovation of,

xvi–xvii, 150, 151–59
Cocke, Jim, 252
Compton, Jerry, 119–20, 123
Conley, Russ, 48
Convy, Bert, 194
coordinated command, 35–36, 63–65, 85–87, 88
Cornyn, John, 126
Cory, Martin, 45, 46
Cowboy's, 105
Crimmins, Patrick, 17–21, 37, 52–54, 65–66, 73, 248. *See also Midland Reporter-Telegram*
crisis, reactions to, 99–100
Cronkite, Walter, xiii, 158
Czech, Richard: and arguments over leadership, 62; and Chip McClure's arrival on scene, 70; and city police administration reform, 56–57; on difficulty of rescue, 66; and media coverage of rescue, 81; and movie about rescue, 203, 208, 210; and planning of rescue, 145; post-rescue life of, 246–47; on progress of rescue, 73, 90; on PTSD and O'Donnell, 225; reaction to Baby Jessica news, 57–58

Dallas Morning News, xiv, 160, 163
daycare licensing, 163–64

INDEX

Dellwood Mall, 52–53
Delta 191 disaster, 99
Dickens, Greg, 44
Dickie, Tony, 109
Donaldson, Sam, 158, 159–60
Donaldson, Vaughn, xv, xvi, 3–4, 223–24
Dorlay, Wayne, 48
Dorr, Karen Danaher, 199
drilling, 58–61, 73, 82–93, 101–2, 146–48, 163, 179, 183
Dunagan, Lloyd, 16

Easter, 143–44
emergency response: arrival of, 31–39; calls for, 16, 19; exhaustion of rescuers, 90–91, 100–101; and reactions to crisis, 99–100. *See also* rescue of Baby Jessica
Everybody's Baby: The Rescue of Jessica McClure (1989), 207–10, 247. *See also* movie deal
exhaustion, of rescuers, 90–91, 100–101
explosive ordnance disposal training, 78–79
Eyre, David, 184, 186, 207, 208–9, 229

Fahd, King (Saudi Arabia), 29
Farish, Teresa, 250
Federal Express, 147–48, 212–13
"feeding the beast," 164
Felice, Dave, 35–37, 73, 87–88, 189, 248
FerreTronics, 25, 26
Fields, Chris, 9
firefighters: pranks among, 225–27. *See also* Forbes, Steve; Midland Fire Department; O'Donnell, Robert
Firefighting Resources of Southern California Organized for Potential Emergencies (FIRESCOPE), 65
First National Bank, 25–26, 30
Fiscus, Kathy, 164
Fletcher, Scott, 58
Forbes, Steve: attention given to, 187, 192–94; and extraction of Jessica from well, 145, 174, 176, 177, 178, 180–81, 182, 185, 191; hears news about Jessica, 92; and plaque commemorating rescue, 252; post-rescue life of, 238, 248–49; on rescue, 190
Foster, Dave, xvi
Foster, John, 51, 184, 186
Frandsen, Olaf, 245–46
Freedom of Information Act (FOIA), 125–26
Free Will Baptist Church, 133, 134
Furgeson, Royal, 79

INDEX

Furnad, Bob, 159, 162

Gaylon, David, 47
Gideon, Wayne, 57
Gilbert (Hurricane), 77
Giles, Geneva, 70
Gilmore, Laura, 107–8
Gilvey, Denise E., 204–5
Glasscock, Andy: affair of, 110–11, 115, 129; arrest and charges of, 167–71; arrives on scene, 32, 33, 34; attends O'Donnell's funeral, 233; attention given to, 75–80, 187; and Bynum's mental health struggles, 108–10; and Chip McClure's arrival on scene, 71; confirms Jessica's consciousness, 82–83; explosive ordnance disposal training of, 78–79; and extraction of Jessica from well, 182, 184; following rescue, 189; and KMID's advantage in news coverage, 52; as liaison for McClure family, 75–76; and movie about rescue, 206, 209–10; MPD accusations against, 122; MPD discipline of, 122–25; and news crews' arrival on scene, 37, 38; on Oprah Winfrey's visit, 198; rests during rescue, 100–101; sentencing and imprisonment of, 237–38, 241–44; on Spivey, 203; and temperature control in pipe, 71–73, 81
Glasscock, David, 77
Glasscock, Jennifer, 180
Glasscock, Lynne, 76–78, 79, 110, 111, 124, 125, 167, 168–71, 189
Glasscock, Michael, 167, 168, 180
Grant, Alice, 179–80
Graves, Debbie, 163–64
Green, Cindy, 136, 137
Green Dragon, 85, 90, 163, 183, 184. *See also* rathole rig
Griffith, Mary R., 252

Haile, Jeff, 204
Hall, Bobbie Jo "B. J.," 31, 32–33
Hall, Mike, 250
Hallmark, Paul, 96–97
Halo above the City (Farish), 250
HBO, 153
Heidelberg, Nathan, 104
Higgins, Darla, 251
Highland Communications Group, 202–3, 206
Hill, Gayle, 174
Hillrichs, Julie, 212
Hollan, Susan, 163
Hollandsworth, Skip, 27–29
Holman, Dave, 92
Holt, Bobby, 214, 215
Houchins, Eddy, 104–6

Houston Post, 183
Huber, Phil, 37, 38, 52
Hubley, Whip, 209, 210
Huntsville Morning News, 18
Hurricane Gilbert, 77
Hutchinson, Alan, 14
hydro drill, 146–48, 179
hypothermia, 71–73, 81

incident command systems, 35–36, 63–65, 85–87, 88
industrial environments: accidents in, 63–64; considerations for rescue missions in, 62
Interscope Communications, 205, 206–7

jackhammers, 101–2
Jauz, Tina, 114, 118, 119, 124–25, 128
"Jessica, Baby." *See* McClure, "Baby Jessica"
Johnson, Laurie, 211
Johnson, Tom, 154, 155, 159
Johnson, Wayne, 84
Jones, Bill, 145–48, 179
journalism: and aftermath of rescue, 191–92; arrival of news crews on scene, 37–38; attention given to Glasscock, 75–80, 187; attention given to O'Donnell, 187, 191–95, 238–39; attention given to rescuers, 187, 191–95; author's occupation in, xiv–xvi; on Baby Jessica story, 18–21, 51–56, 81, 98–99, 134–35, 148–50, 174–77, 179–80; cable news, xvii, 151–53; CNN's news-reporting innovation, xvi–xvii, 150, 151–59; Crimmins's coverage of Baby Jessica story, 52–54; Crimmins's start in, 18; culture of metropolitan versus small-town, 20; and extraction of Jessica from well, 182–84, 185, 186; first printed news on Baby Jessica story, 53–54; live television coverage of Baby Jessica story, 55–56; live television news coverage in Permian Basin, 51–52; localizing national stories, 164–65; McClure family dealings with news media, 148–50; and post-rescue treatment of Jessica, 217; PTSD from, 99–100; satellite news coverage of rescue, 98–99. *See also Midland Reporter-Telegram*
Junell, Robert, 237, 242

Kander, John, 207
Karr, Kenny, 149–50
KCRS-AM, 160–61

INDEX

Keller, Lawana, 17
Kier, Mary Alice, 199
King, Larry L., 11, 27
King Consultants, 7–8
Klatt, Eddie, 47
Klunick, Chip, 180
KMID-TV, 51–52, 55, 174–75, 191, 213
Knight, Keshia, 212
Koppel, Ted, 184
Kraft, Bernard, 108
KTLA, 165

Lamar Elementary, 14–15
LaMotte, Greg, 162–63, 177
law enforcement officers: dangers facing, 104; police reform under Czech, 56–57; shooting of, 103–4. *See also* Bynum, Paula; Glasscock, Andy
Leon, Scott, 159
Lewis, Holden, 134–35
life support, 73
lights, 89
Lilly, Dave: actions following rescue, 187; arrives on scene, 101; backlash against, 229; breakthrough into well casing, 173, 174; drills in tunnel, 145; and extraction of Jessica from well, 180–81; interviewed by Oprah Winfrey, 198; and movie about rescue, 209; recruited for rescue, 95, 96–97
Lilly, Doris, 95, 187
live television news coverage, xvi–xvii, 51–52, 55–56
Livermore, Norman B., 65
localization, of national news stories, 164–65
Lofland, Chuck, 183
Lovejoy, Clint, 105, 109, 119–20
Lucas No. 1 Spindletop well, 22
Lunsford, Ann DeLong, xii–xiii
Lunsford, Donnie, xii
Lunsford, Doug, 132, 133
Lunsford, Dutch, 131, 132–34, 161
Lunsford, Margie, 131, 134, 161

Malone, Jess, 167, 168, 170
Maple, Earl, 157, 180, 184
Marcus, David, 160
Marcy, Randolph B., 12
Martin, Hartwell, 143, 144
Martin, Joanne, 140–41, 143, 192, 231, 238
Massengill, Chester, 84
Maxwell, Jane, 159
Maxwell, Phil, 77
McClure, "Baby Jessica": falls down well, 15, 16–17; following rescue, 189–95; gifts and donations for, 212–13; Glasscock as liaison for, 75–77; injuries of, 213–14; post-extraction care of,

185–86; post-rescue life of, 250–52; post-rescue treatment of, 211–19; throws out first pitch for Texas Rangers, 76–77; trust fund for, 251–52. *See also* rescue of Baby Jessica

McClure, Chip: arrives on scene, 70–71; background of, 66–67; and Bushes' visit to Jessica in hospital, 215–16; Bynum referenced in book of, 104; childhood near-death experience of, 69–70; dealings with news media, 149–50; donations given to, 213; and extraction of Jessica from well, 177, 182; hears news about Jessica, 67–68; on hydro drill, 146–47; and movie about rescue, 206, 210; post-rescue life of, 249–50; rumors following, 250

McClure, Reba "Cissy": and arrival of first responders, 31, 33–34; attends O'Donnell's funeral, 233; background of, 67; dealings with news media, 148–49; depiction in print news, 53, 54; discovers Jessica in well, 15, 16–17; and extraction of Jessica from well, 176, 177, 182; following rescue, 189; on Jessica's trust fund, 251–52; and movie about rescue, 206, 210; post-rescue life of, 249; and post-rescue treatment of Jessica, 211, 214; relationship with Robbie O'Donnell, 235; on selling story, 201; sings to Jessica, 82; speaks with Reagans, 215

McClure, Rod, 69

McClure Rescuers Association, 202–6

McCoul, Chip, 35–37

McGowan, Jim, 47

McGraw, Robyn, 212

Menchaca, Rick, 119

Mendenhall, Charlie, 202

Mercantile Texas Corporation, 30

MGM/UA Television Productions, 201–3

Midland, Texas: author moves to, xiv; clings to limelight, 218; culture and history of, 11–14; failure of First National Bank, 25–26, 30; as located in Bible Belt, 131; news coverage in, 41; oil industry in, 21–22, 23, 26–28; Oprah Winfrey's visit to, 197–99; water of, 16; and World War II, 131–32

Midland Airfield, 132, 213

Midland Fire Department, 86–87; Central Fire Station,

265

135–36, 225–27; Station 2, 35–36; Station 8, 44–48. *See also* firefighters
Midland Memorial Hospital, 211–19
Midland Police Department (MPD). *See* Bynum, Paula; Glasscock, Andy; law enforcement officers
Midland Reporter-Telegram: author's employment with, xiv–xvi; and Bynum's self-shooting, 111–29; continuing coverage of rescue, 148–49. *See also* Crimmins, Patrick
mining, 93, 95. *See also* Lilly, Dave
Minton, Meta, 128–29
Modisett, Kimberly, 212
Moore, James, 71
Moore, Jamie, 16, 71
Morales, Daniel, 251
Morgan, Tom, 171
movie deal, 197, 199–210, 228. See also *Everybody's Baby: The Rescue of Jessica McClure* (1989)
Murrah Federal Building, 8–9, 223

Nabisco Brands, Inc., 200
National Enquirer, 149, 201
National Law Enforcement Officers Memorial Fund (NLEOMF), 103
New York Stock Exchange, 214–15
Norden, Carl, 132
Nye, Ramona, 19–20, 148, 160

Oakland, California, wildfires, 64–65
Odessa, Texas, 11–12, 23, 41
O'Donnell, Casey, 142, 143–44, 231, 232–33, 234, 241, 249
O'Donnell, Chance, 98, 142, 143, 231–33, 241, 249
O'Donnell, Ricky, 4, 44, 139, 195, 221–22
O'Donnell, Robbie: Easter celebration with family, 143, 144; on *Everybody's Baby*, 210; post-rescue life of, 235, 249; and Robert's celebrity status, 193; and Robert's involvement in rescue, 98; and Robert's post-rescue return, 238; and Robert's prescription drug dependence, 221–23; and Robert's suicide, 221, 231–35
O'Donnell, Robert: attention given to, 187, 191–95, 238–39; background of, 144–45; backlash against, 228–30; celebrates Easter with family, 143–44; death of, 3, 4–5,

42–43, 221, 231–35, 241; employment of, 7–8, 43–44; and extraction of Jessica from well, 145–46, 173–74, 176, 177–79, 180–82, 185, 186, 191; final hours of, 139–41; following rescue, 189–90, 238–39; interviewed by Oprah Winfrey, 198; and McClure Rescuers Association, 203; mental struggles of, xvi, 43–44, 80, 140–41; and movie about rescue, 199–201, 204, 206, 207, 208–9, 210; and Oklahoma City bombing, 8–9, 41–42; and plaque commemorating rescue, 252; prescription drug dependence of, 44–49, 141, 221–24; volunteers for rescue effort, 97–98
Ohlmeyer, Donald W., 200
Ohlmeyer Communications Company (OCC), 199–200
oil industry, 13–14, 21–30, 91
Oklahoma City bombing (1995), 8–9, 41–42, 139
Olkewicz, Walter, 209–10
O'Neill, Tex, 193
Oprah Winfrey Show, 197–99, 229
Orlean, Susan, 13
Orozco, Trey, 48
Ott, Gary, xvi, 20–21, 128

overexertion, 63
Owens, Tim, 31–32, 45

Painter, Gary, 77, 213, 233
Partridge, Toby, 191, 211
Pena, Victor, 47
Penn Square Bank, 29
Permian Basin, 13–14, 21–25, 51–52, 60, 62
Permian Estates, 15–16, 131
Perry, Dave, 91, 92
Petree, Jerry, 36–37
Phillips, Clinton, 64
Phillips, Jim, 92
Poe, David, 7, 42
Poe, Yvonne, 9, 42–43, 140
police. *See* law enforcement officers
pornography: and arrest of Andy Glasscock, 169–71; and sentencing of Andy Glasscock, 237
post-traumatic stress disorder (PTSD), xv, xvi, 3–4, 5, 8–9, 99–100
Powell, Ferrell, 212
Preston Drive, xi–xii
Price, Aubrey, 93, 95–97
Pruden, David, 46
Pulitzer Prize, 245–46

Queen, Bill, 186, 215
Quinn, Matt, 135

Ramirez, Grace, 179
Rampi, Alfredo, 164
ranchers / ranching: early, 12–13; modern, 94
Rather, Dan, 182
rathole rig, 58–61, 66, 73, 82, 83–85. *See also* Green Dragon
Reagan, Nancy, 197, 214–15
Reagan, Ronald, 197, 214–15, 229
Reese, Debbie, 211
religiosity, 132–34
Republic of Texas separatists, 79
rescue of Baby Jessica: aftermath of, 189–95; arrival of emergency responders, 31–39; attention given to Glasscock, 75, 79–80; author's recollections concerning, xi–xiv; breakthrough into well casing, 173–74; call for more drillers, 91–92; calls for emergency responders, 16, 19; chance for potential accidents in, 63–64; Chip McClure arrives on scene, 70–71; Chip McClure first hears news about, 67–68; and CNN's news-reporting innovation, 151–59; completion of parallel shaft, 82; Crimmins's reporting on, 52–53; difficulty of, 66; disorganization at scene of, 62–63, 65, 85–87, 88, 101–2; drilling, 58–61, 73, 82–93, 101–2, 146–48, 163, 179, 183; early approach to, 38–39; exhaustion of rescuers, 90–91, 100–101; extraction of Baby Jessica, 175–79, 180–86; fifteen-year anniversary story of, xiv–xvi; first printed news reporting on, 53–54; hydro drill brought in, 146–48, 179; Jessica's responses to rescuers, 82–83; life support, 73; Lilly recruited for, 93–97; live television news coverage of, 55–56; and Midland's clinging to limelight, 218; movie about, 197, 199–210, 228; news coverage of, 65–66, 134–35, 148–50, 160–63, 174–77, 179–80; O'Donnell's inability to let go of, 80; O'Donnell's involvement in, 97–98, 140–41; Oprah Winfrey dedicates episode to, 197–99; and passage of time, 89–90; plans for extraction, 38–39, 61; plaque commemorating, 252; post-extraction care of Jessica, 185–86; rights to story of, 199–203; satellite news coverage of, 98–99; temperature control, 71–73, 81; tension in, 99–100;

unorthodox ideas for, 135–37, 146, 180. *See also* emergency response

Roberts, James: and arguments over leadership, 62; and attention given to rescuers, 192–93; and extraction of Jessica from well, 182, 184; and extraction plans, 61, 81–82; following rescue, 189–90; and media coverage of rescue, 81; and movie about rescue, 203, 208, 210; and O'Donnell, 47, 233, 235; and planning of rescue, 145; post-rescue life of, 247; reaction to Baby Jessica news, 56, 57–58

Robinson, Kragg, 58, 59–61, 73, 81–82, 84–85, 203

Robinson, Oscar, 82

Rogers, Ken, 108, 109

Rolls Royce dealership, 13

Russert, Tim, 182

Sabreliner 40A, 93–94, 96

Saldono, Theresa, 201

San Angelo Standard Times, 19

San Marino, California, 164–65

Satcom 1, 153

Saudi Arabia, 29

Schonfeld, Maurice "Reese," 151, 154

Schorre, Al, 119–20

Scout Sunday, 132–33

Serviatus, Jim, 53, 245–46

Sevey, Jim, 109

Shanks, Gary, 106

Shaw, Bernard, 157–58, 159–60

Shaw, Scott, 176, 183, 185, 186, 244–46

Short, Ronald, 146, 180

Shusman, Steve, 155–56, 159

Smith, Darrell, 205

"Smith, Sandy Lou," 168–70, 242

Spence, Charlie, 128, 203

Spencer, Jerry D., 3, 4, 5

Spivey, Loehr, 202–3, 206, 207

Sprague, Maxine, xv, 16–17, 161

Sprague, Raymond, 16, 47

Stanley, Tabitha, 107

Stanton, Robert, 183

Starks, Pete, 17

stock market drop, 214–15

Stringer, Howard, 154

temperature control, 71–73, 81

Texas Department of Health and Human Services, 163

Texas Rangers, 76–77

Texas Water Commission, 15

Thames, Willie, 191

Third Degree, 193–94

Thomas, Carroll, 201, 207

Thompson, Ronald G., 243

time, passage of, 89–90

Todd, Ed, 147, 217, 218

trust fund, 251–52

INDEX

Tryon, Randel, 112–13
Tubb, Terry, 217, 219
Turner, Ed, 159
Turner, Ted, 151–55, 156–57, 159–60, 177
Turner Communications Group, 152
20/20, 192

UHF deregulation, 152–53
Urby, John, 106, 107, 108–9, 111, 113, 114, 116, 118–19, 123–24, 128–29
Utlee, Gary, 162

Vermicino, Italy, 164
Viney, Shelton, 199, 216, 218
Vogel, Robert A., 44, 47
volunteers, 62, 63, 64, 85–86, 89, 101–2, 173, 197–98
Von Fremd, Mike, 184
VonHolle, Jeff, 112

Walker, Bill, 31–32, 34–35, 83
wealth: in Midland, 13, 22; in oil industry, 25, 28
Weaver, Chad, 126
well(s): description of, 15; Lucas No. 1 Spindletop well, 22; in Permian Estates area, 15–16. *See also* rescue of Baby Jessica
West Side Baptist Church, 133
West Texas: history of, 11–13; journalism in, 164; oil industry in, 13–14, 21–29, 91; spot news and viewership in, 37. *See also* Midland, Texas; Odessa, Texas
Westar I satellite, 153
WFAA, 98–99, 135
whistleblowers, 121–22
White, Jim, 54–55, 114, 149, 174, 175, 179, 201, 207
White, Robert, 35–36, 48, 56
Wilcott, Curt, 34, 53
wildfires, 64–65
Wiley, Bill, 45, 46
Williams, Clayton, 93–95
Williams, Dean, 73
Williams, Don, 167
Williams, Jack "Doctor Jack," 135–36, 137, 170, 227–28, 230
Winfrey, Oprah, 197–99, 229
Wood, Rick, xvi, 183–84, 186
World War II, 131–32
WTCG, 152, 153
Wunsch, Rodney, 55, 174–75, 177, 191

Yeakey, Terrance, 9
Yamani, Sheik Ahmed Zaki, 29
Younger, Charles, 199, 214, 216, 217, 218–19

Zoellers, Barbara, 133
Zoellers, Eugene "Gene," 133–34

Printed in the USA
CPSIA information can be obtained
at www.ICGtesting.com
CBHW010054210924
14733CB00006BA/525